SpringerBriefs in Entrepreneurship and Innovation

Series Editors
David B. Audretsch
Institute of Development Strategies, Indiana University,
Bloomington, Indiana, USA

Albert N. Link
Department of Economics, University of North Carolina Greensboro,
Greensboro, North Carolina, USA

T0155764

For further volumes:
http://www.springer.com/series/11653

Barry Bozeman • Craig Boardman

Research Collaboration and Team Science

A State-of-the-Art Review and Agenda

 Springer

Barry Bozeman
Center for Organization
    Research and Design
Arizona State University School
    of Public Affairs
Phoenix, Arizona, USA

Craig Boardman
John Glenn School of Public Affairs
The Ohio State University
Columbus, Ohio, USA

ISSN 2195-5816   ISSN 2195-5824 (electronic)
ISBN 978-3-319-06467-3   ISBN 978-3-319-06468-0 (eBook)
DOI 10.1007/978-3-319-06468-0
Springer Cham Heidelberg New York Dordrecht London

Library of Congress Control Number: 2014938624

Printed on acid-free paper

Springer is part of Springer Science+Business Media (www.springer.com)

Acknowledgements

The authors gratefully acknowledge the assistance of Andrew Kao, Arizona State University, in acquisition and organization of literature reviewed here and Mary O'Brien, also Arizona State University, for her help in manuscript preparation.

Acknowledgements

Contents

1 **Assessing Research Collaboration Studies: A Framework for Analysis** 1
  1.1 The Collaboration Imperative 1
  1.2 Research Collaboration Concepts 2
  1.3 Knowledge-Based and Property-Based Collaborations 3
  1.4 Focus and Boundaries 4
  1.5 Questions Guiding this Study 5
  1.6 Empirical Focus 5
  1.7 Theory Framework: Scientific and Technical Human Capital 6
  1.8 Collaboration and S&T Human Capital 7
  1.9 A "Logic Model" Approach to Organizing Relevant Literature 8
  1.10 Decision Rules for Selecting Articles 10

2 **Inputs, Resources and Research Collaboration** 13
  2.1 Individuals and Groups: STHC as Inputs and Resources 13
    2.1.1 Formal Training at the Individual and Project, Organizational Levels 13
    2.1.2 Past Productivity 15
    2.1.3 Social Capital 15
    2.1.4 Career Status and Past Career Experiences 17
    2.1.5 Motivations and Related Characteristics 18
  2.2 Materiel as Inputs and Resources for Team Science and Research Centers 21
    2.2.1 Tangible Capital and Labor 21
    2.2.2 Prior Knowledge and Technology (Field/Industry Level) 22
    2.2.3 Organizational Capital 23

3 **Processes and Activities in Research Collaboration** 27
  3.1 Project Level Management and Leadership 27
    3.1.1 Management and Leadership at the Project/Team Level 27

        3.1.2 Strategic Management and Leadership at the
               Organizational Level 30

4 **The Outputs, Outcomes and Impacts of Research Collaboration** 33
    4.1   Introduction 33
    4.2   Outputs and Impacts for "Knowledge-Focused" Collaborations 34
    4.3   Outputs and Impacts for "Product-Focused" Collaborations 36
    4.4   Outputs and Impacts for Scientific and Technical Human
          Capital Impacts 39
    4.5   Negative Impacts of Boundary-Spanning Research
          Collaboration 41
          4.5.1 Possible Negative Effects on Teaching and Education 41
          4.5.2 Possible Negative Effects on Research
                 and Research Careers 42
          4.5.3 Negative Impacts on Industry 45

5 **Effectiveness Questions and Research Recommendations** 47
    5.1   Pondering the Effectiveness Questions 47
          5.1.1 Research Management Approaches 47
          5.1.2 Learning from Failure 49
          5.1.3 Intellectual Property 49
    5.2   Recommendations for New Research 50
          5.2.1 Recommendation 1: Meta-Choice 51
          5.2.2 Recommendation 2: Research on Institutional
                 Failures and the "Dark Side" of Collaboration 51
          5.2.3 Recommendation 3: Scientific and Technical
                 Human Capital 52
          5.2.4 Recommendation 4: Focus on Management of
                 Large-Scale University-Based Research Centers 52
          5.2.5 Recommendation 5: Pursue Field Experiments 53
          5.2.6 Recommendation 6: More Impact-Focused
                 Research; Integrated Collaborative Teams
                 to Study Collaboration 54

References 57

Chapter 1
Assessing Research Collaboration Studies: A Framework for Analysis

1.1 The Collaboration Imperative

Today in most science, technology, engineering and mathematics (hereafter STEM) fields more than 90 % of research studies and publications are collaborative (Bozeman and Corley 2004), leading to a "collaboration imperative." Not only does team-based collaborative research more often lead to high impact research and to commercial uses of research as reflected in patents (Wuchty et al. 2007), in many fields it is not possible to thrive as a single investigator. If one's work depends on access to samples or specimens or to extremely expensive shared equipment, then collaboration and research are essentially one in the same, and, thus, the collaboration imperative. Thus, despite significant variation by field, discipline and geography (De Stefano et al. 2013), contemporary STEM research is dominated by collaboration, teams, networks and co-authorship.

The past few decades' trends in increased research collaboration are owing to a variety of factors, including, among others, the rapid specialization of science and the melding of fields (Porter and Rafols 2009; Haustein et al. 2011), new enabling collaborative technologies (Zhou et al. 2012; Muscio and Pozzali 2012; Tacke 2011), resource sharing imperatives (Neveda et al. 1999; Ynalvez and Shrum 2011), and new public policies explicitly encouraging collaboration (Ponomariov and Boardman 2010; Wallerstein and Duran 2010; Van Rijnsoever and Hessels 2011).

Beyond simply setting context for our analysis, we are not particularly concerned with either the amount of research collaboration or even the broad institutional factors conducing collaboration. The increase in team projects and in individual-level and institutional level research collaboration is well documented (for an overview see Bozeman et al. 2013) and, thus, we take the proliferation of collaboration as a given, focusing instead on the processes and "mechanics" of collaboration as well as determinants of collaboration effectiveness.

We begin with the most with a fundamental question that should drive any research assessment: what is the nature of the "dependent variable(s)?" Below, we present an overview of various concepts of research collaboration.

B. Bozeman, C. Boardman, *Research Collaboration and Team Science,* SpringerBriefs in Entrepreneurship and Innovation, DOI 10.1007/978-3-319-06468-0_1, © The Author 2014

1.2   Research Collaboration Concepts

The term "research collaboration" is used to describe relationships between individuals but also relationships between organizations and between individuals with organizations. Of course, when organizations collaborate, it is actually individuals who are relating to one another, thus the distinction is not so clear as it might seem. One source of ambiguity in individual research collaboration is that some collaborators think of their collaborative activities almost as essentially coterminous with co-authorship whereas others take a broader view. For this reason, and also because co-authorship is conveniently measured, much of the published work about research collaboration focuses on co-authorship. As Katz and Martin (1997) point out, co-authorship is at best a partial indicator of collaboration.

For present purposes, we define collaboration as "*social processes whereby human beings pool their experience, knowledge and social skills with the objective of producing new knowledge, including knowledge as embedded in technology.*" By this definition, collaboration need not be focused on publishing articles and, indeed, collaborations often are more concerned with technology development, software or patents and may have no publication objective at any point. Increasingly, large teams of specialists interact to produce research and technology and, in some cases at least, some of the collaborators never meet or even interact with one another. Still, this seems to us to qualify as a collaboration since it is a bringing together of talents for the purpose of knowledge creation and usually results in an identifiable knowledge product (e.g. scientific paper, patent).

In our definition, collaboration is chiefly about scientific and technical human capital (Dietz et al. 2001), a notion that we outline in detail below and a model that frames our review and analysis. For the present, suffice to say that scientific and technical human capital is the amalgam of the researcher's social capital (network ties and linkages) and her human capital (native abilities as they are developed through education and training). Obviously, organizations and institutions as well as financial and physical resources are vital to the success of many research collaborations, but our definition suggests that one who *only* provides resources does not qualify as a collaborator. Thus, a person who has knowledge of laboratory equipment may bring that form of human capital to a relationship and, by our definition, be a collaborator, whether or not recognized as a co-author or whether or not assigned any patent rights. A researcher might be listed as a co-author because she has provided material or performed an assay (Stokes and Hartley 1989). In some cases an individual may make a major contribution to research and neither obtain nor desire co-author credit. For example, a mentor may help shape a vital part of a doctoral student's dissertation, perhaps even providing the core idea. Such a relationship can be a true collaboration.

Broader notions of collaboration are not often easy to measure. Focusing on co-authorship alleviates many measurement problems and, thus, many useful studies (e.g. Heffner 1981; Vinkler 1993; Wagner 2005; Heinze and Bauer 2007; Mattsson et al. 2008) of research collaboration begin and end with the co-authored publication. Despite the challenges of research with a broader concept of research collaboration, several studies do examine collaborations with measures seeking to tap more than co-authorship. Most of these studies (e.g. Melin 2000; Bozeman and Corley 2004; Bozeman and Gaughan 2010) rely on researchers to nominate their collaborators or provide a broader definition of collaboration (Jeong et al. 2011) and then develop indicators based on the broader definition.

1.3   Knowledge-Based and Property-Based Collaborations

We consider two different types of R&D productivity here, each quite relevant, even crucial, to research collaboration. Increments to knowledge are generally measured in terms of scientific and technical articles produced, cited or, more rarely, demonstrably used. Increments to property and wealth are typically measured in terms of patents, new technology, new business start-ups, and, more rarely, profits. We use the term "knowledge-based collaborations" and "property-based collaborations" to distinguish the two, while at the same time recognizing that the categories are not mutually exclusive.

Research results show clearly that knowledge-based research collaboration tends to enhance STEM research productivity (Pravdić and Oluić-Vuković 1986; Lee and Bozeman 2005; Wuchty et al. 2007; Huang and Lin 2010). The contribution of research collaboration to knowledge has been examined in studies employing a wide array of methods, quite different data and focused on different organizational and sector environments have provided convergent evidence of beneficial impacts.

In the case of property-based collaboration's effects on profits, wealth and economic development, the models tend to be particularly complex, but support the idea that collaboration is beneficial to commercial product and process development (Franklin et al. 2001; Shane 2004; Dietz and Bozeman 2005; Link and Siegel 2005; Perkmann and Walsh 2009). Such research is vital inasmuch as most research collaborations occur within and among industrial firms simply because that is where most research occurs. However, most studies of collaboration do not focus on industrial firms, except perhaps in partnership. One reason is that firms often are not anxious for others to study them, especially when the firms have trade secrets or proprietary knowledge or technology at stake. Moreover, research tells us that the objectives, composition, and content of research in industry tend to be quite different from those found in universities, government or nongovernmental organizations (NGOs) (Crow and Bozeman 1998; Cohen et al. 2002; Guellec and Van Pottelsberghe de la Potterie 2004).

Since those researching collaboration are to a very large extent located in universities and since university research activity tends to be more transparent, the preponderance of research on collaboration focuses on collaborations among university scientists or collaborations between university scientists and industrial or government scientists. This leads us to consider the boundaries for the present study, as presented in the next sub-section.

1.4   Focus and Boundaries

While the *study* of scientific collaboration certainly has not kept pace with the *practice* of collaboration, the increase in studies of the topic has nonetheless been impressive. However, studies of scientific collaboration and team science are quite diverse with respect to level of analysis, study approach and method, sectoral focus, and institutional setting. Since the majority of persons performing and publishing research on scientific collaboration are located in universities, it is perhaps unsurprising that studies of university researchers' collaborations comprise a substantial component of the literature. While our study will draw from some of the university collaboration literature, our focus is on "boundary-spanning research collaborations," specifically:

1. University-industry partnerships and
2. Industry interdisciplinary research collaborations.

While we do not wish to go into great detail about the considerable differences in organizational designs for university-industry institutions and industry-based collaborative institutions, each of these has a considerable accompanying literature, but at least a few words are in order. With respect to university-based organizations, we have elsewhere (Bozeman and Boarman 2003, 2004) described in detail the idea of a multi-discipline, multi-purpose university research center (MMURC), research centers that have more than one university affiliate, more than one discipline and have multiple functions including not only research but also education and outreach and, often, technology development. Most, but not all of these MMURCs are tasked to build relations and collaborate with industry. The National Science Foundation's Engineering Research Centers and Science and Technology Centers typify such organizations but many states' governments have developed centers of excellence programs providing quite similar research organizations. In some cases universities themselves have developed complex research centers that meet the structural requirements of MMURCs. Much of the collaborative research with industry is provided through MMURC's but this does not describe the university of university research centers (see Bozeman and Boardman 2013 for a more complete and data-grounded taxonomy) and, moreover, a significant amount of collaboration between university scientists and engineers and industry occurs through much less complex mechanisms.

With respect to multi- and inter-firm collaboration there is, likewise, a diverse array of institutions and collaborative arrangements. In interfirm R&D alliances there is often a formal contractual basis where firm A pays firm B to conduct R&D, and in some cases collaborative R&D. These are quite different than industry research consortia, which are typically much more broad-based and typically include more than a few firms. Firms also engage in joint venture collaborations of a variety of sorts with partners playing any of a wide variety of roles. The term "research partnership" is a quite general label that could apply to almost any collaborative arrangement. For more detail on the various modes of inter-firm R&D collaboration types and institutions see the literature assessment provided by Hagedoorn et al. (2000).

1.5   Questions Guiding this Study

For the most part we will be focusing here on research collaborations that occur in within the institutions that have been developed to facilitate collaboration rather than, say, inter-firm collaboration based on contracting. The policy and strategy implications are more pressing with respect to the more complex, newer and more innovative approaches to collaboration rather than, say, contract R&D or ad hoc collaborations.

Given the above institutional focus, taken with our assessment or the most pressing questions of research collaboration, our study considers the following questions:

- How do different management approaches and leadership styles influence the effectiveness of science teams? For example, different approaches to establishing work roles and routines and to the division of labor may influence team effectiveness.
- How do current tenure and promotion policies acknowledge and provide incentives to academic researchers who engage in team science?
- What factors influence the productivity and effectiveness of research organizations that conduct and support team science, such as research centers and institutes? How do such organizational factors as human resource policies and practices and cyber infrastructure affect team and science?
- What types of organizational structures, policies, practices and resources are needed to promote effective team science, in academic institutions, research centers, industry, and other settings?
- What does the available research on university-industry research partnerships and within-industry team science tell us about effective research management approaches and partnership models?
- How do intellectual property and conflict of interest concerns affect the collaborative processes and scientific and translational outcomes?

1.6   Empirical Focus

As we demonstrate below, for some of the above questions there is an abundant research literature and for others only a scant literature. The "size of the draw" for this synthesizing study is further governed by the fact that we focus almost exclusively on studies providing *empirical evidence*. Our definition of empirical evidence is a broad one that includes not only studies based on aggregate statistics and statistical models, but also systematic qualitative studies and case studies. However, even with this expansive concept of empirical evidence, this focus eliminates from our consideration a significant proportion of the research on university-industry partnerships and industry interdisciplinary research collaborations. We do not consider literature based solely on conceptual models, unverifiable personal insights, anecdotes or opinions that are not supported by data.

By using the search filters "university-industry partnerships" and "industry interdisciplinary research collaborations" and adding the additional filter "empirical evidence" we eliminate a significant proportion of the literature on research collaboration and team science. Moreover, we believe that research in other settings may in some instances be relevant either because the ideas transcend sectors and institutions or, more often, because the help pose questions (even with answering them) relevant to university-industry partnerships and industry interdisciplinary research collaboration. Thus, we employ outside-the-analytical-boundary studies in some instance below, but we are in every case cautious about the limitations of context.

After providing an overview of relevant research, organized according to a systematic categorization, we turn our attention to drawing lessons and providing recommendations. Based on the review of research we provide and assessment of the state of research and significant gaps in research, but the chief focus is to use existing evidence-based research to suggest possible approaches to improve understanding and perhaps performance of university-industry research partnerships and industry interdisciplinary research collaborations.

1.7   Theory Framework: Scientific and Technical Human Capital

Syntheses and assessments of the literature require no particular theory framework and in some instances can rely on the extant literature to suggest theory limits or even coherent theory perspectives. But in this case we feel that the body of theory we have referred to elsewhere (e.g. Bozeman et al. 2001; Bozeman and Corley 2004; Gaughan and Robin 2004; Edler et al. 2011) as *scientific and technical human capital theory* adds some value for a study that seeks not only to synthesize literature but also to suggest its utility (e.g. Shrivastava and Mitroff 1984).

Scientific and technical human capital (STHC) is not a novel or an original theory but an amalgamation of two bodies of researcher and theory that have long been prominent in studies science and technology policy and management. STHC is the sum of scientific and technical and social knowledge, skills and resources embodied in a particular individual (Bozeman et al. 2001). It is both human capital endowments, such as formal education and training, and social relations and network ties that bind scientists and the users of science together (Bozeman and Rogers 2001). STHC includes the human capital is the unique set of resources the individual brings to his or her own work and to collaborative efforts. Generally, human capital models (Becker, 1962; Schultz, 1963) have developed separately from social capital models (Bourdieu 1986; Bourdieu and Wacquant 1992; Coleman 1988, 1990), but in the practice of science and the career growth of scientists, the two are not easily disentangled.

Scientific networks are crucial to scientists throughout the professional life cycle because these networks consistently affect scientific work and even the ability to obtain work. The networks include not only scientists themselves, but also a variety of other actors who use and enable science, including funding agents, vendors, entrepreneurs, equipment developers, technicians, and public officials, among others. One early statement (Dietz et al. 2001) provides a summary of STHC:

- Science, technology, innovation, and the commercial and social value produced by these activities depends upon the conjoining of equipment, material resources (including funding), organizational and institutional arrangements for work and the unique S&T human capital embodied in individuals.
- While the production function of groups is not purely an additive function of the S&T human capital and attendant non-unique elements (e.g., equipment), it resembles closely an additive function. (The "missing ingredient" in such aggregation is the quality of the fit of the elements to the production objectives at hand.)
- At any level, from the individual scientist to the discipline, field or network, the key focus is on capacity and capabilities. What factors enhance capacity, diminish it or simply change the reservoir of capabilities inherent in individuals and groups?

1.8   Collaboration and S&T Human Capital

Scientific collaboration often plays a critical role in developing STHC, especially in those cases where the collaboration takes on mentoring characteristics, such as when a more experienced scientist collaborates with a junior scientist, a post-doctoral researcher, or a graduate student. But STHC also is enhanced by peer collaboration, especially when those involved are from different organizations, fields or sectors. In such cases, diverse collaborators bring together quite different knowledge, experiences and social ties and exchange these, enhancing the cooperating organization or team's fund of STHC but also each of the participants'.

In general, STCH focuses on developing capacity- in the individual, the team, the organization and even the scientific field. Thus, the STHC model defines capacity in terms of the assets individuals bring to science and technology problems, including not only their formal learning, training, experiential and tacit knowledge, intelligence and native abilities (i.e. human capital broadly defined), but their social networks and social ties, including their linkages to institutions that produce, consume and disseminate knowledge. Most important, for present purposes, *we assume that research collaborations are undertaken to enhance the STHC brought to science and technology problems and challenges.*

We view STHC theory as an especially useful perspective for articulating the logic of research collaborations and therefore for organizing the literature on the topic. As we see it there are two key reasons STHC theory is most appropriate (and one of these reasons is not that we ourselves developed STHC theory). First, STHC theory nicely captures both the institutional and resource-based views of collaboration. STHC theory is "institutional" to the extent that it characterizes individuals' and organizations' professional linkages and network ties to institutions with distinctive norms and expectations for individual and organizational outcomes (e.g., Ponomariov 2013; Jonkers and Cruz-Castro in press). STHC theory is "resource-based" to the extent that it professional linkages and network ties can vary based on different sorts of organizations and organizational partnerships such as across private companies and government centers programs sponsoring university research centers (e.g. Murray 2010; Rogers 2012; Menzies 2012).

The second (and more important) reason we use STHC theory for analyzing the literature is that it is much more specifically focused on "additionality" (Aerts and Schmidt 2008) in science teams and research centers.[1] STHC theory's emphasis on *capacity for change* (versus on institutional norms and/or resources only) as an explanatory mechanism is well aligned with the intentions of the government, academic, and industry decision makers who collectively establish and participate in boundary-spanning research collaborations. Accordingly, STHC theory provides an interpretive frame relevant analysis for research evidence use in decision-making.

It is important to note that we do not seek here to develop or extend STHC theory but rather to use its basic assumptions as a guiding perspective. This is especially important in a study where the "dependent variable" has no necessary or direct bearing on such relevant productivity indicators as income, profit, profit maximization or any such monetized outcome. While such indicators are relevant to many collaborations, especially intra-firm research and collaboration among units with a firm or a complex for-profit entity, they have less value for collaborations among institutional sectors or among government, nonprofit and business entities. A focus on STHC gives attention to outcomes valued by all stakeholders in collaborations. While the use of STHC assumptions colors the entire paper, in the concluding section we come back to implications of the STHC to interpretation of the chief findings.

1.9 A "Logic Model" Approach to Organizing Relevant Literature

Having set boundaries for the study and decided upon a broad interpretive model, the challenge remains as to the best approach to choosing among the vast number of "empirically-based studies of industry-based research collaboration and university-industry collaboration." The organizational approach we take here is in some respects similar to the "logic model." Logic models are structured frameworks, often used as a program evaluation or project management tools, are structured frame-

---

[1] This refers to assessing empirically whether or not the outcomes of boundary-spanning research collaborations would have occurred even if the collaboration had not ever occurred. It is analogous to the more general empirical challenge of discerning empirically the net effects of any public policy, management technique, or treatment/intervention when randomization is not a research design possibility. Bozeman and colleagues (2013) discuss this problem, which the science policy literature refers to as the problem of "additionality," at length in their recent review of the literature on individual-level research collaboration.

works for relating resources, inputs, outputs and impacts (for detailed explanation see Julian 1997; McLaughlin and Jordan 1999; Knowlton and Phillips 2012). While most such models are not inordinately complex and often include simplifying assumptions, logic models do have the advantage of imposing a certain discipline in causal thinking (Kellogg Foundation 2001). Our approach here is not exactly an application of a logic model but it is a structured attempt to relate inputs (independent variables) to outputs and impacts, assuring some clarity of causal reasoning (though not necessarily validity of causal inference, a much more difficult and fragile analytical problem). Logical models are not unlike contingency models in organization theory (e.g. Shaw et al. 2011; Van de Ven et al. 2013)- they demand focus on leverage points.

One advantage of a structured contingency or logic model approach is that it encourages a "theory of change" (McLaughlin and Jordan 1999), especially relevant for the study of team science and research centers[2] involving diverse sets of actors (e.g., researchers, clinicians, external stakeholders) from diverse institutions and with diverse incentives and goals. The logic model approaches can entail a set of systematic "if-then" statements that can explain when boundary-spanning research collaborations like team science, research centers, and industry consortia are successful and when they are not. The second advantage is that the logic model approach as a method will help to identify *important* knowledge gaps in the literature, ones that require attention.

Most adherents of a logic model encourage the sure use of a guiding substantive models or conceptual theory (McLaughlin and Jordan 1999). As indicated above, we use scientific and technical human capital (STHC) theory, which typically is applied at the individual level of analysis and defines STHC as the sum total of a researcher's professional network ties, formal training and technical skills, tacit knowledge, and resources broadly defined (Bozeman et al. 2013). But this approach can be readily applied at the organizational and project levels if the definition of STHC is expanded to include not just the totality of STHC represented by a team's or research center's scientists and engineers, but also to include organizational level resources and network ties, e.g., the social structures and norms to facilitate coordination and communication amongst the diverse sets of researchers as well as the management and leadership strategies and practices that the social structures and norms for coordinating diverse sets of actors imply (Barney 1995).

Our model for organizing the literature on boundary-spanning research collaborations is depicted in Table 1.1 below.

The attributes for the first two sections (columns) of the logic model were chosen by applying the basic definition of STHC at both the individual and organizational levels. The outputs and outcomes addressed in the third and fourth columns of the logic model were selected not by our broadened definition of STHC per se, but by applying the key series of propositions of STHC theory, adapted to team science, research centers, industry consortia, and the like. The propositions go as follows.

---

[2] Though typically these terms are used to refer just to boundary-spanning research collaborations in a university and/or government setting, the current review addresses fully the literature on boundary-spanning research collaboration amongst private firms.

Table 1.1 Logic model/framework for organizing the empirical literature on research collaboration

Inputs and resources	Processes and activities	Outputs, outcomes, and impacts
People and groups	Project level management and leadership	Enhanced knowledge-focused output and impacts
Formal training	Project management and leadership	Enhanced property-focused outputs and impacts
Past productivity	Collaborative research, development, and testing	Enhanced scientific and technical human capital (short run outputs)
Social capital	Organization level strategic management	Negative impacts of collaboration
Career status and experiences	Strategic management and leadership	
Motivations and related characteristics	Marketing for additional inputs and resources	
Materiel		
Capital and labor		
Prior knowledge and technology		
Organizational capital		
Formal		
Informal		

First, the establishment of team science and research centers convenes new sets of human and social capital. Second, this new convention requires management and leadership at multiple levels (e.g., individual, project, organizational) to facilitate communication and coordination within each new set of human and social capital. Third, each new set over time will beget new human capital at the individual, project, and organizational levels *that would not have occurred otherwise*, due to formal and tacit learning amongst the researchers within each set. Last, the new human capital at all levels will lead to intended and unintended outcomes and impacts *that would not have occurred otherwise*.

1.10   Decision Rules for Selecting Articles

We apply three guiding (i.e. flexible) decision rules for selecting articles for the review. These rules are depicted in Table 1.2. The first rule addresses whether a focal article is evidence-based (in the sense of the term described above). If the focal article is not evidence-based, it is excluded from the review. For example, there are a number of very good perspectives pieces on the need for more (or less) research collaboration between universities and industry, but they are excluded because they do not present a clearly articulated research question involving use of quantitative or qualitative data.

If the focal article "passes" the first decision rule, the next decision rule is whether the article addresses research collaboration per se, either by way of including data

Table 1.2 Decision rules for including and organizing articles using the framework

Is the focal article evidence based?
If yes, include
If not, exclude
Does the focal article address research collaboration per se?
If yes, include
If not, exclude
Does the focal article address attributes under more than one component of the logic model?
If yes, which components?
If yes, how will the discussion of the focal article be different across the different components?
If no, what is the attribute and which component does it fall under?

for variables measuring collaboration inputs, processes, outputs, and/or outcomes or by sample selection (e.g., sampling researchers affiliated with NSF centers). This rule results in the exclusion of a number of popular empirical articles examining the connection between government support for university research and scientific productivity. However, the rule also implies that the subset of these studies that addresses empirically policies and institutions designed to facilitate boundary-spanning research collaborations (e.g., CRADAs, university research centers) *are* included in the review.

If the focal article "passes" the second decision rule, the last rule is determining what part of the guiding logic model the article fits. This rule is not a vital once since ours is a guiding model for organizing the literature and, of course, various research findings stand on the own regardless of our collaboration. Still, the model provides convenience. If the article examines empirically one or more of the attributes from the framework as an input, then that article is addressed under "inputs and resources." In turn, if the same focal article models empirically yet another attribute from the logic model as an outcome, then that attribute is *not* addressed in the "inputs and resources" section (beyond brief mention and a cross-reference), but rather is discussed in the "outcomes and impacts" section of the review.

Unfortunately this final decision rule cannot be implemented based on the actual research design of the focal article. A large proportion of the literature on boundary-spanning research collaborations does not employ quasi-experimental designs (i.e., it lacks panel data and comparison groups) and therefore cannot establish causal direction empirically. Therefore from a research design perspective for many articles it is impossible to know whether a particular attribute is, empirically, an input or output and thusly it is difficult to (again from a research design perspective) validly and reliably decide what section(s) of the logic model an article should be included in. One this account, we simply make do when causality is obscure.

We have provided in this chapter a summary of the analytical tools we use for our study, including a definition of research collaboration, the categorization of research collaboration as knowledge-focused vs. property-focused, the decision rules for determining which literature to consider, the STHC analytical framework and, of course, the research questions we set for ourselves. In the next chapter we begin the first aspect of the review, focusing on inputs and resources and their effects in research collaboration.

Chapter 2
Inputs, Resources and Research Collaboration

Here we examine resources and inputs into collaboration as they interact with the institutional settings for collaboration, including teams, research centers, and firms, among others. We examine (1) people (researchers and research support) with their respective "caches" of human capital, (2) materiel both physical and intangible, and (3) social structures and norms.

2.1 Individuals and Groups: STHC as Inputs and Resources

This component of the framework addresses the attributes of individual and interactive groups of researchers, specifically their STHC, including (1) formal training, (2) past productivity, (3) social capital, (4) career status and career experiences, (5) personal characteristics, and (6) organizational characteristics and motivations. Though STHC theory proposes that these are interrelated, for convenience in description we treat them separately in our review of the literature. Moreover, most published articles focus on just one of these aspects of human capital. Most of the articles discussed in this section are at the individual level of analysis, though a few consider team and organizational level characteristics (as teams and organizations are comprised of people and too can exhibit behaviors like productivity and other characteristics like motivations).

2.1.1 Formal Training at the Individual and Project, Organizational Levels

Formal training is perhaps the most traditional way to operationalize human capital. For the most part the economics literature uses the concept to distinguish the college-educated from non-college educated. But this sort of variation seldom exists in boundary-spanning research collaborations wherein almost all researchers have doctorates.

B. Bozeman, C. Boardman, *Research Collaboration and Team Science,* SpringerBriefs in Entrepreneurship and Innovation, DOI 10.1007/978-3-319-06468-0_2, © The Author 2014

The literature on boundary-spanning research collaborations distinguishes different backgrounds and formal training at the doctoral level, albeit more typically as control variables rather than as antecedents of primary (or even secondary) interest. Most of this "controlling" has occurred with the work of Bozeman and colleagues and their series of survey- and curriculum vitae-based research on the scientific values of and participation in research collaborations by academic researchers in the US. This series of articles (e.g. Bozeman and Corley 2004; Bozeman and Gaughan 2007, 2011; Boardman and Ponomariov 2007; Boardman and Corley 2008; Bozeman and Boardman 2013) has elicited no consistent results connecting formal training (which Bozeman and colleagues mostly operationalize as respondents' PhD field) to particular research collaboration strategies or behaviors. Perhaps the most consistent result across this series of articles is that there are considerable differences across fields with regard to the propensity to collaborate with industry and most find that engineering disciplines are the most likely to collaborate with industry.

A more useful operationalization of formal training for designing and implementing team science and research centers is at the project and/or organizational level, specifically addressing the heterogeneity of disciplines for a particular project or organization. The reason this is more useful is because both public and private sector research is becoming increasingly multidisciplinary due to the increasing complexity of the scientific and technical innovation required to address social, economic, health, energy, defense, and other national problems (Zerhouni 2003). Also, methodological and epistemic norms across disciplines can be quite different, sometimes diverging sharply (Snow 1964; Clark 1983; Becher 1989; Kekale 2002; Turner et al. 2002; Van Gigch 2002a, b) and therefore an impediment to boundary-spanning research collaboration (Goldman 1986, Corley et al. 2006).

There are a few studies that look at what we prefer to call "disciplinary heterogeneity" (to avoid unnecessary discussion, *here*, of the differences between interdisciplinary, multidisciplinary, and transdisciplinary research). Chompalov et al. (2002) find as disciplinary heterogeneity increases, so does productivity, but also so does heterogeneity of incentives and motives to collaborate and thusly collaborations become more hierarchical as well as more formalized organizationally.

Many other studies focus on the relationship between collaboration and disciplinary differences or collaboration and multi- and interdisciplinary issues. Some studies (e.g. Qin et al. 1997; Bordons et al. 1999; Schummer 2004; van Rijnsoever and Hessels 2011), measuring the degree of interdisciplinary collaboration and drawing conclusion based on those differences). Other studies are more normative, evaluating or prescribing means to overcome or exploit disciplinary differences (e.g. Hall et al. 2008; Porter et al. 2006). The lesson from all these studies is that multi- and inter-disciplinary collaboration often prove difficult to organize and manage but also that they are especially likely to prove beneficial in terms of both the creation and diffusion of knowledge and technology.

2.1.2 Past Productivity

Human capital constitutes a vital transfer mechanism for knowledge and technologies across boundaries, e.g., from universities to industry, from one academic discipline to another, from one individual to another (Schartinger et al. 2001). Due to easy availability of productivity data for some industries, a popular way to measure "observable" human capital is to assess productivity, which according to economists may be considered proxies for human capital to the extent they represent productivity an individual and/or organization would not be able to produce without a particular human capital component (Aboud et al. 2005). However, for the research collaboration literature there are not many studies addressing bibliometric and/or patenting productivity as an input *per se* (i.e., as a key predictor rather than as an antecedent control variable) into team science or research center processes, outputs, and/or outcomes. One example of this sort of study (i.e., conceptualizing scientific and technical productivity as an antecedent to research collaboration) are those studies focused on "star scientists," i.e., scientists who are very productive bibliometrically because of their own innovative techniques and ideas (e.g., Zucker and Darby 1996). However, this particular "brand" of scientist is demonstrated to collaborate less to protect her or his techniques and ideas from imitation and emulation (Zucker and Darby 1996). Another example are articles focused on predicting and explaining the bibliometric productivity of team science or research centers that include lagged measures of bibliometric productivity as antecedent control variables for contemporaneous bibliometric productivity (e.g., Ponomariov and Boardman 2010).

One reason the research collaboration literature has not followed many economists in operationalizing past productivity as contemporaneous human capital is that seldom do these two literatures meet. Another reason is that bibliometric and patenting productivity have generated more interest in policy and scholarly circles as outputs and outcomes of research collaboration. A final reason may be that past research productivity may not be a valid measure of either inputs or outcomes for team science and research centers. Cronin (2001) and Garg and Padhi (2001) for example call into question the measurement validity of research productivity as a proxies for both inputs into and outputs of collaborative research as it may occur in team science or research centers due to what they term "hypercoauthorship" wherein not all coauthors contributed expertise or human capital per se to the collaborative project and resultant article, but rather are listed as coauthors for reasons other than expertise/human capital contributions.

2.1.3 Social Capital

According to the STHC model (Bozeman et al. 2001)[1], human capital is much more than formal training or, just discussed, past research productivity, but rather is the

---

[1] There is to date no singular measure or scale for the scientific and technical human capital idea (and this was not the original intention of Bozeman and colleagues 2001), so the different aspects of the idea are addressed as separate human capital attributes.

summation of a number of individual attributes and experiences, including but not limited to social capital and the human capital associated with that social capital. Though there is a vast literature on social capital and its variable impacts, only the studies on the role(s) of social capital in boundary-spanning collaborations are addressed here. Because of the smallness of the literature on the role(s) of social capital specifically in boundary-spanning research collaborations, some of the general studies on social capital in collaborations generally are also included.

Much of the empirical work in this area emphasizes prior acquaintance and trust as an important input into research collaborations. Prior acquaintance as an input into new team science or research centers can enhance collaboration from the start due to decreased if not avoided transaction costs associated with engendering trust amongst collaborators from different disciplines, universities, firms, government agencies, sectors, and even age cohorts (Granovetter 1985, 1992; Marsden 1981; Williamson 1975; Commons 1970). One of the key findings from the general empirical literature on boundary-spanning collaboration (not necessarily research collaboration) is that acquaintance and the trust it engenders leads to progressively fewer and less formal structures and authorities for governing the collaboration (Gulati 1995).

Among the more interesting empirical studies of the role of trust in boundary-spanning research collaborations is one focused on inter-firm research collaboration. Dodgson (1993) addresses trust not at the interpersonal level due to the frequency of turnover in high technology companies (especially during the "tech bubble" when this study was conducted. Instead, Dodgson emphasizes interorganizational (versus interpersonal) trust characterized in two case companies by what he calls communities of interest, organizational cultures receptive to external inputs, and frequent and egalitarian communication of information about the status and purpose of the collaboration. This is perhaps one of the more interesting studies on trust and research collaboration not just because of its emphasis at the organizational level of analysis and its focus on firms (which in theory need trust less due to their ease abilities to formalize research collaboration relations relative to academic researchers and government agencies), but additionally because organizational culture is the most difficult aspect of organizations to cultivate much less change (Tolbert and Hall 2010).

Many other studies at the individual level of analysis show trust to correlate positively and significantly with collaboration processes, outputs, and outcomes (Sonnenwald 2007; Creamer 2004; Hara et al. 2003; Sonnenwald 2003; McLaughlin and Sonnenwald 2005; Knorr-Cetina 1999; Krige 1993). Trust has been a big emphasis in studies of research collaboration because without it collaboration in most cases does not happen (Olson and Olson 2002). Though their data are at the individual level of analysis, perhaps most relevant of these articles to the current study are the studies by Shrum et al. (2007) and by Zucker et al. (1995), both who show that in team science and research centers trust is most readily engendered amongst same-discipline and same-institution researchers in multi-discipline, multi-institution research collaborations. These findings point up all the more the importance of understanding how to develop a cultural "fit" at the organizational level in team science and research centers. Indeed this problem has seen as much research on problems in boundary-spanning research collaborations as on success (Boardman and Gray 2011; Gray et al. 2013).

When there is not prior acquaintance and trust, it can be very important to start the strategic planning process well before a team science or research center begins operations, i.e., during the planning phase. To avoid high transaction costs due to lack of acquaintance and trust (Kogut 1998), researchers and stakeholders present at the beginning of a boundary-spanning research collaboration can lower these costs and increase trust and acquaintance by collectively discussing and selecting additional researchers with whom to collaborate (Lipnack and Stamps 1994; Scharpf 1978; Landau 1991). These processes as they pertain to team science and research centers are addressed by way of review of many team science and research centers articles that address processes and activities empirically or with case study.

2.1.4 Career Status and Past Career Experiences

Career status and career experiences have been used as proxies for individuals' levels of expertise or human capital in a number of studies. In their recent review of university-based research collaborations with industry Bozeman and colleagues (Bozeman et al. 2013) use age as a proxy for past experience at research collaboration and suggest that it is perhaps "one of the most apparent personal factors one might expect to have an effect on collaboration." But they do not consider age explicitly as a proxy for human capital. Instead they suggest that age may have an impact on collaboration frequency and intensity because the longer one conducts research, the more collaborators she or he acquires.

The findings for age with regard to research collaboration processes, outputs, and performance are mixed. Ponomariov and Boardman (2010) find no statistically significant relationship between age and the rate or intensity of academic researchers collaborating with industrial researchers (the authors address university-industry collaboration using a variety of co-authorship measures). But there is a selection issue here insofar that academic researchers who work with and sometimes co-author with industrial researchers are not the norm and represent but a small proportion of academic researchers overall.

Other studies find age correlating positively with commercial and/or collaborative research activities and also with more traditional academic activities like publishing in the open literature. Haeussler and Colyvas (2011) find increased commercial outputs on the part of older academic researchers. Boardman and Ponomariov (2007) find similar results with regard to academic researchers' tenure status and their respective valuation of commercially relevant versus traditional academic research. Both studies suggest under the traditional academic reward template emphasizing peer reviewed publication in the open literature that the opportunity cost of engaging in more commercial research collaborations is lower for older versus younger university faculty. Both studies also suggest that not until researchers have had enough experience and productivity will they be sought out by companies, which implies a human capital dimension/interpretation of researchers' age.

Past career experiences have also been shown to have an impact on research collaboration and scientific and technical productivity. Rijnsoever and Hessels (2011) find heterogeneous research experiences to correlate positively with both disciplinary and interdisciplinary research and, similar, Aschoff and Grimpe (2011) show academic researchers who work early on in their careers with industry to be more involved with and publish more with industry based researchers later on. Ponomariov and Boardman (2010) show the rate and intensity of both interdisciplinary and university-industry co-authorship to correlate positively with academic researchers' affiliation with university-industry research centers, using a panel of bibliometric data for 57 academic researchers over the entirety of their respective academic careers. Boardman and Ponomariov (2013) use comparative case study for a purposive sample of NSF university-industry research centers to suggest (anecdotally, N=21) that past career and educational experiences in management may have an impact on how formalized and horizontally differentiated research centers are.

Other studies assess career experiences using both more and more sophisticated constructs. Dietz and Bozeman (2005) emphasize changes in job sectors throughout scientific and engineering careers and how these affect productivity in terms of both publications and patents. Using the curriculum vitae of 1,200 scientists and engineers from the USPTO (US Patent and Trademark Office) database, the authors show that diversity of career experiences leads to diversity in scientific and technical productivity. Specifically, the authors show that inter- and intra-sector job changes (i.e., university to industry and vice versa, university to university, industry to industry) leads to higher publishing and patenting rates. The authors use STHC theory to explain these findings: career transitions of academics involving collaborations with different disciplines and with industry enhance human capital and therefore productivity.

2.1.5   Motivations and Related Characteristics

In the numerous studies surveying researchers at the individual level, many control for motivations and related characteristics such as the age, gender, and ethnicity of respondents. Melin (2000) suggests that there are many personal reasons to engage in research collaborations like team science and research centers (and also acknowledges exogenous reasons as well) but emphasizes the social aspects of collaboration as a primary motivation at the individual level of analysis. Beyond the social aspects, amongst samples of academic researchers the job security that comes with tenure has been shown to affect attitudes towards collaboration. Boardman and Ponomariov (2007) find that not having tenure negatively and significantly correlates with a willingness to work with industry among a stratified random sample of academic researchers affiliated with NSF Engineering Research Centers. Using a broader data set on a national sample of academics whether they are affiliated with a research center or not, Corley and Bozeman (2004) find that tenure status does not correlate significantly with personal collaboration strategies. This contrast speaks to the selection effect of surveying just academic researchers who self-select into research centers like ERCs. A final influential motive for academic researchers joining collaborative endeavors like team science and research centers is job satisfaction (Lee and Bozeman 2005; Coberly and Gray 2013).

Age is more appropriately in our view discussed as a proxy for career status and longevity of career experiences and is discussed above (see 2.1.3). Gender and ethnicity have been shown to be an important personal collaborator attributes in the scientific community. There has been a lot of focus on women and ethnic minorities in science over the past two or so decades (especially at the NSF) due to their under-representation of women and minorities in academic science (Pollak and Niemann 1998; Johnson and Bozeman 2012).

The impact of gender on research collaboration activities and outcomes can be direct or an interaction with other individual characteristics like tenure status. Bozeman and Corley (2004) show that women tend to collaborate more than men do in academic science. And this finding holds for women at all stages of their careers (e.g., tenured or not, research group leader or not) and for different types of academic careers in science (tenure track versus research faculty). The authors also find that non-tenure track women are more likely to collaborate with other women than with their male counterparts. In a more recent yet similar study Bozeman and Gaughan (2011) seek to "break" previous models suggesting that gender matters to collaboration and specifically boundary-spanning collaboration. In this attempt they developed a new nationwide (US) survey data set including over 1,700 respondents and weighted by field and gender with an especial focus across the survey items on research collaborations with private companies and the motivations for entering these collaborations. The study showed that females collaborate more and in more ways with industry and was the first to demonstrate that the "gender effect" holds when controlling for numerous other personal attributes. Published almost simultaneously, Rijnsoever and Hessles (2011) find very similar results regarding women and the rate and intensity of interdisciplinary research collaborations (i.e., being more likely for women than men).

In contrast, the numerous studies using the "industry involvement index" (see Bozeman and Gaughan 2007) for a detailed explanation of the index, which is a weighted gradient) finds gender to have no statistically significant correlation with academic researchers' collaboration intensity with industry (Bozeman and Gaughan 2007; Gaughan and Corley 2010; Ponomariov and Boardman 2008, 2010; Boardman 2010).

Ethnic minority status has also been a factor in empirical and case based research on team science and research centers. Sonnenwald (2007) provides an extensive review not only of the relationship between participation of ethnic minorities in research collaborations but also addresses field work for alleviating some of the mistrust, misunderstanding, and conflict that can arise. In this sense minority status is yet another "boundary" that is spanned in boundary-spanning research collaborations, and like the usual suspects (i.e., disciplinary boundaries, economic/sectoral boundaries, institutional boundaries) if not managed properly these can lead to role conflict and role strain (Boardman and Bozeman 2007), distrust (Fisher and Ball 2003; Secrest et al. 2004), and a lack of informal mechanisms for coordinating diverse sets of actors like goal congruence (Boardman 2012). See Sonnenwald (2007) for a more extensive review.

Practically all of the research on the motivations and related characteristics for organizations entering research collaborations focus on private firms. A predominance of these studies provides a resource-based explanation of one sort or another (e.g., size in terms of employees and/or R&D budget). Others emphasize government incentives, geography, and leadership. The findings are quite consistent perhaps because most of these studies address private firms whose organizational environments and therefore stakeholder sets and goals are more uniform when compared to university research centers and team science.

Organizational size (usually measured as both number of employees and proportion of budget allocated to research and development) is the most frequently cited characteristic explaining which firms join industry R&D consortia and/or research collaborations with other firms. Practically all of this research finds a positive correlation between size and motivation (e.g., Angel 2002; Bayona et al. 2001; Fritsch and Lukas 2001; Santoro and Chakrabarti 2002; Kaiser 2002). But some studies show relatively unique findings. Kleinknect and Reijnen (1992) find size to decrease rather than increase the likelihood of Dutch firms collaborating with one another on research and development. Aloysius (2002) finds that firms of comparable sizes are most likely to enter into formal research collaborations with one another.

Last, some of these findings for size and motivation for firms to collaborate with one another in research may be spurious. For example, Santoro and Chakrabarti (2002) conclude that larger firms participate in university research centers to build new research capacity outside core research areas whereas smaller firms participate in centers to fulfill core research areas. But the sample of firms they use participate in different types of university research centers with very different capacities for "radical" research and development (Ettlie and Ettlie 1984; Damanpour 1996) deviating from existing knowledge and technology versus "incremental" research and development (Ettlie and Ettlie 1984; Damanpour 1996) building predominantly on existing knowledge and technology. Though Santoro and Chakrabarti acknowledge that "unique structural and contractual features distinguish university research centers" (p. 1164) and include in their analysis firms participating in centers both with and without the support of National Science Foundation centers programs, they do not emphasize theoretically or empirically the programmatic status of the centers included in the analysis, but rather focuses on firm size.

Other motivations for firms to collaborate in research and development include having common precompetitive research challenges (Ouchi and Bolton 1988; Greis et al. 1995; Sakakibara 1997; Katila and Mang 2003; Mowery et al. 1998; Hayton et al. 2013) and, related, weak competition from other firms (Sakakibara 2002); spatial proximity (Fritsch 2001) and, related, being located in a large urban area (Angel 2002); having an internal champion of inter-firm research collaboration who acts as a gatekeeper to identify firms with which to collaborate in research and development (Fritsch and Lukas 2001, Mathews 2002), as well as incentives for inter-firm research collaboration from government (Sakakibara 2001; Hayashi 2003) and, related, national culture (Steensma et al. 2000).

2.2 Materiel as Inputs and Resources for Team Science and Research Centers

This component of the framework addresses the attributes that individual researchers, including their (1) tangible capital such as equipment and infrastructure, and labor such as post-docs and graduate students and (2) prior knowledge and art including past research and technology.

2.2.1 Tangible Capital and Labor

The literature on tangible capital for university research centers and government and university team science is sparse. Most of the work emphasizing tangible capital as inputs is from the literature on firm-firm research collaborations and, to a lesser extent, on industry consortia. In our view, the lack of emphasis in the literature on tangible capital in university research centers and university and government team science is due to the transition over the past few decades of what constitutes competitive advantage in scientific and technical innovation at the university and national levels. However, a few studies focused on experimentation in high-energy particle physics emphasize tangible capital in research centers and team science that are not firm-based (e.g., Krige 1993; Galison 1997).

Focusing predominantly on firm-firm research collaborations, Hagerdoorn et al. (2000) observe the popularity of resource dependence perspective among strategic management scholars in explaining boundary-spanning research collaborations amongst private firms, which collaborate on research projects to gain access to resources and capabilities that enable them to develop and sustain competitive advantages. Others also focusing on firms (Becker and Peters 1998; Camagni 1993) emphasize the sharing of resources to reduce uncertainty and to realize cost savings as well as economies of scale and scope. Audretsch et al. (2002) count firms' network ties to universities as a tangible resource though they are really discussing the human capital in universities as much as the research laboratories and university research centers with complementary infrastructure, equipment, and critical materials.

None of this literature emphasizes access to tangible capital *alone*, but rather emphasizes access to caches or sums of human capital and social capital and labor in addition to more tangible resources like infrastructure, critical materials, equipment, and funds. In other words, whether the theory is called out explicitly or not in the literature on tangible capital in boundary-spanning collaborative research, the STHC approach driving our analysis of the literature in this study is adept for explaining the role of tangible capital in boundary-spanning research collaborations insofar that the STHC theory implies a number of important *int*angible processes and activities as well as other types of inputs and resources, e.g., human capital, without which tangible capital cannot be implemented.

The small literature on labor in boundary-spanning research collaborations emphasizes labor as a standalone input (rather than part of a broader cache of resources) and focuses on graduate students in university-industry research collaborations. This literature is bifurcated, with some authors suggesting that graduate students' involvement with industry is beneficial to technology transfer (Ponomariov 2009; Bozeman and Boardman 2013) and also to education (Bozeman and Boardman 2013), though other authors characterizing such graduate student involvement with industry as disruptive and potentially harmful to traditional graduate education (e.g., Slaughter et al. 2002; Slaughter and Rhoades 2004).

Though we treat them as one literature here, really the two perspectives constitute separate literatures. Seldom does empirical study emphasizing graduate students roles in technology transfer address outcomes related to education, like teaching and student support, and almost as a rule does the *discourse* (typically not empirical research) (Slaughter et al. 2002; Slaughter and Rhoades 2004) on the potentially disruptive and harmful nature of student involvement with industry go unconcerned with the potential benefits of this involvement for education as well as for economic outcomes. While both sets of results (i.e., graduate students correlating positively with technology transfer in university-industry research collaborations, graduate students experiencing negative unintended consequences) can be correct in isolation, Bozeman and Boardman (2013) address both types of outcomes to demonstrate academic researchers who work with different types of university-industry research centers also to mentor and teach more graduate students. Behrens and Gray (2001) finds a null relationship comparing graduate students working in university-industry research centers to their counterparts not working in these centers or comparable arrangements. We address these studies further in Sect. 2.3 in under the "enhanced scientific and technical human capital" subheading.

2.2.2 Prior Knowledge and Technology (Field/Industry Level)

The literature on prior knowledge and technology at the field level is one that has no focused on boundary-spanning research collaborations directly, but no less it is quite important, as the state of a particular field of inquiry can have substantial impacts on the very establishment (Bozeman and Boardman 2003) and also the organization and management of such collaborations (Boardman 2012). Conceptually, prior knowledge and technology at the field level can be "radical," i.e., relatively divergent from existing knowledge and practice or it can be "incremental," i.e., focused on the application of existing knowledge and practice (Ettlie et al. 1984). There are just a few studies that address the radical-incremental dichotomy as it relates to boundary-spanning research collaborations. Damanpour (1996) demonstrates collaborations focused on research and development in radical fields to have more complex organizational structures (e.g., centralized decision making, formal contracts, rules of use) than firms focused on incremental research and development. Boardman (2012) similarly finds across different types of university-industry research centers radical research and development to be more conducive to structured management practices than to incremental research and development (Damanpour's and Boardman's findings regarding organization-level structure are discussed further in Sect. 2.3).

2.2.3   Organizational Capital

This component of the framework addresses the attributes that boundary-spanning research collaborations require to facilitate intended activities, outputs, and outcomes. Here we define organizational capital broadly as anything that can be implemented at the collaboration design and implementation stages (or potentially in a redesign stage after a formative evaluation) to induce coordinated problem solving amongst the people and materiel addressed in sections above. Our review does not address exogenous organizational capital, such as political support and popular support, insofar that these are much less manipulable by policy and program decision makers in government and industry.

We consider organizational capital to be perhaps the most important resource and input into boundary-spanning research collaborations in terms of competitive advantage in scientific and technical innovation.[2] Research centers and team science *that are not firm based* will not be able to enhance STHC and achieve "additionality" (see Bozeman et al. 2013) outcomes without organizational capital because it is fundamental to the processes and activities of research centers and team science. Yet there is not much literature directly focused on organizational capital in boundary-spanning research collaborations that are not firm based because they are predominantly informal and more difficult to observe (Hagedoorn et al. 2002). There is research on formal contracts for firm-firm research collaborations and memoranda of understanding for industry consortia, but these mechanisms become difficult to emulate when government agencies and universities are the "home" contexts for the collaboration (Bozeman and Boardman 2003).

We do not address the research on organizational capital in firm-based research collaborations here. The reason for this is that the literature on industry research consortia and firm-firm research collaborations using formal contracts and memoranda of understanding does not address variation in organizational capital per se, but rather uses formal organizational capital as a sample selection device, i.e., the

---

[2] Increasingly there is equity across research universities in the US and more broadly internationally across developed nations (at least those with explicit science and technology policies) in terms of expertise, capital such as equipment and infrastructure, funding (though at the national level the US and China eclipse other nations), human capital (see sect. 2.1 above), labor.

samples are comprised either of all memoranda or all contract-based collaborations. The literature on the productivity and impacts of industry consortia and firm-firm collaborations in research and development is appropriately addressed in sect. 2.5 alone. Also, in this section we do not address the management practices and leadership styles that are a function of organizational capital. These are addressed in the next section on "activities and processes." Here informal and formal structures are considered as inputs rather than processes when they are operationalized as organizational features not as behaviors on the part of managers or leaders in boundary-spanning research collaborations.

Organizational capital can be formal or informal (Barney 1995) in research centers and team science when universities and government agencies rather than firms are the "home" institutions. Most of the time, it seems these endeavors use informal coordination mechanisms to facilitate coordinated problem solving within diverse sets of scientists and engineers. This (reliance on informal organizational mechanisms) is the case typically when there are multiple institutions (Chompalov et al. 2002), there is a long history of working together among the researchers (Galison and Hevly 1992; Krige 1993; Knorr Cetina 1999), and when there is trust (Gulati 1995), as well as inertia (Li and Rowley 2002).

One of the few studies that address organizational capital in university-industry research collaborations is a recent study by Boardman (2012). He uses as series of 21 cases of NSF university-industry research centers to describe (not predict and explain, this is a qualitative investigation) which centers rely on informal mechanisms and which rely on formal ones to coordinate diverse sets of scientists and engineers. Boardman controls for many of the other inputs addressed above, including multidisciplinarity, size, radical versus incremental research and development, and the past career experiences of the center directors to find a substitution effect of formal for informal mechanisms. When informal mechanisms for coordinated problem solving such as resource interdependence and goal congruence are not there, the director implements formal structures to facilitate coordinated problem solving and vice versa.

In a recent edited volume on cooperative research centers by Gray et al. (2013), two chapters address organizational capital in university-industry research centers that are university-based. Garrett-Jones et al. (2013) use survey data to demonstrate that formal structures and authorities can mitigate role conflict, which is interpretable as an outcome of a lack of goal congruence and resource interdependence (i.e., the informal coordination mechanisms Boardman 2012 addresses). Similar, Davis and colleagues find that informal relations between center directors and university administrators correlates with center performance, perhaps due to resource support due to legitimacy.

Chompalov et al. (2002) identify informal organizational capital as fundamental to boundary-spanning research collaborations. Though they do not use the same terminology (the term "organizational capital" is borrowed from the strategic management literature), the authors use the same theoretical premise (i.e., transaction costs economics as applied to organizations in Ouchi (1980)). Although they develop a typology of organizational structure for research collaborations, they argue (and find) that hierarchy is not a defining characteristic (in this they were responding to the literature by Krige on high energy particle physics experiments), but rather that the defining characteristic is consensus, which could occur for a number of reasons, including the informal mechanisms identified by Boardman (2012, discussed above), i.e., goal congruence and resource interdependence.

In our previous reviews of the literature on research collaboration, trust is emphasized as organizational capital (e.g., Boardman and Bozeman 2006; Bozeman et al. 2013), at least to the extent that it is discussed alongside some of the articles we discuss here focused on formal and informal mechanisms for coordinating diverse actors. Because STHC (scientific and technical human capital) theory is driving the logic of this literature review, in contrast to our past efforts we discuss the literature on trust as social capital, e.g., prior acquaintance. No matter how this subset of the literature (on trust) is organized, an interesting yet unaddressed question is whether informal organizational capital (e.g., goal congruence, resource interdependence) begets trust or vice versa, or both depending on conditions (e.g., prior acquaintance). In our view this is an important dynamic to understand when forming research teams and centers.

Chapter 3
Processes and Activities in Research Collaboration

Here we focus on the basic "factors of production" for boundary-spanning collaborative research and development are put to use by managers and leaders at the project and organizational levels. At the project level we address much of the "science of team science" literature that has recently come into vogue (since about 2008 or so) and that is focused predominantly on NIH supported collaborative research. At the organization level we address much of the "cooperative research centers" literature that is focused on NSF supported collaborative research, especially between universities and industry. Firm-firm research collaboration as well as intra-firm team management research are addressed first in each subsection, as reference points for the team science and research centers literatures, to assess the extent to which these literatures are making general contributions that can *validly* inform decision making for STHC development and deployment with the establishment of new teams and new centers for new problems requiring scientific and technical innovation.

3.1 Project Level Management and Leadership

3.1.1 Management and Leadership at the Project/Team Level

Before getting to the "science of team science" literature, we address the strategic management literature on team/project research collaborations (and collaborations more generally) from management science (which focuses predominantly on firms). Much of this literature emphasizes task characteristics (e.g., complexity), related leadership tactics (e.g., participatory decision making, job rotation), team diversity/heterogeneity, and other project management and leadership activities to induce contributions from diverse actors towards team/project goals.

Starting with task characteristics, a number of studies show a positive correlation between task complexity and team performance (Campion et al. 1993) and some with task interdependence as an important mediating factor for this positive relationship (Campion et al. 1996; Jehn et al. 1999; Wageman 1995). Other studies

B. Bozeman, C. Boardman, *Research Collaboration and Team Science,* SpringerBriefs in Entrepreneurship and Innovation, DOI 10.1007/978-3-319-06468-0_3, © The Author 2014

from the management science literature emphasizing the team level of analysis provide a potential explanation for these relationships (between task characteristics and performance) by focusing on team heterogeneity and resultant management and leadership practices.

First, there is some evidence that heterogeneous teams outperform homogenous teams in terms of institutional and expertise heterogeneity (e.g., Cummings 2004). Jehn (1995), Jehn et al. (1999), and Jehn and Mannix (2001) find among other things that task complexity and interdependence moderate the typically negative impacts, e.g., conflict (Pelled et al. 1999), of team heterogeneity. Different, Earley and Mosakowski (2000) demonstrate with case study and then lab experiments a curvilinear relationship between team heterogeneity and team functioning, with highly homogenous and highly heterogeneous teams outperforming teams with more moderate levels of homogeneity/heterogeneity. Second, there are leadership styles that show evidence of enhancing team performance. Participatory decisionmaking (De Dreu et al. 2001; Edmonson 2003; Kirkman et al. 1999) has been shown to enhance cooperation and performance (usually by mitigating some of the negative aspects of team heterogeneity, like role conflict and goal incongruence (Pelled et al. 1999)). Job design has also been shown to enhance team performance, controlling for task interdependence and complexity as well as team heterogeneity (Campion et al. 1993). Similar to the findings on participatory decision making, Edmonson (2003) shows that leadership in operating room scenarios for cardiac surgery (training for new techniques, not actual surgeries) that demand that clinicians from different disciplines to "speak up" mid surgery (simulation) outperform operating teams without a participatory (and perhaps at times less collegial or at least amicable) leadership style, due to the team level learning that the former facilitates.

According to the strategic management literature focused at the team or project level (not organizational level, which is addressed in below), processes and activities for teams to consider as tools to mitigate conflict and enhance cooperation and performance are general, e.g., leadership for participatory decision making, job design, and task characteristics such as complexity and interdependence, and can be brought about by a number of different practices. In terms of the logic model framing this literature review, these processes and activities are but some tools for developing new STHC from the "factors of production" for collaborative research addressed above and eventually affecting intended outputs and outcomes. What is interesting about the science of team science literature is that many of these factors are constant—e.g., as with most boundary-spanning research collaboration at the project level, team science is heterogeneous and its tasks are complex and interdependent. But the leadership and management practices are not. However, the science of team science literature seems more focused on discrete practices rather than general ones, which when translated to other contexts or teams may not have the same impact.

The science of team science literature is focused predominantly on NIH science teams for addressing diseases and disorders with sets of scientists and clinicians with different backgrounds and training. Much of the early discussion of this literature, by its creators, goes to great lengths to address the differences between cross-disciplinary, multidisciplinary, and interdisciplinary research collaborations (e.g. Stokols et al. 2008; Hall et al. 2008; Borner et al. 2010).

We do not replicate past literature reviews on the science of team science (e.g., Hall et al. 2008; Borner et al. 2010). Instead we discuss broadly the findings from the literature focused on the new population of boundary-spanning research collaborations (i.e., NIH science teams). We do review findings as summarized in these previous reviews and relate these specific findings to more general management and policy mechanisms that decision makers can address no matter how they manifest in a particular instance of team science. We take this approach because in our view the science of team science literature in large part emphasizes complexity rather than parsimony (which is antithetical to the purpose of the current literature review)—by design per the proclamation that the field is in its inductive and therefore taxonomic phase (Hall et al. 2008).

The inductive approach that the science of team science literature has pursued thus far identifies face-to-face proximity, Web-based coordination mechanisms in the absence of face-to-face proximity, shared scientific values (e.g., emphasizing collaborative research across the disciplines), trust, and interpersonal collaboration processes and planning as fundamental to successful team science. The approach has also identified formal structures and authorities and the inclusion of junior faculty without tenure as detrimental to effective team science collaborations. There are two observations to make here per the STHC framework guiding the logic of this review. First, many of these findings are indeed required to render the inputs and resources discussed above as more than the sum of the parts, e.g., for the enhanced STHC and the new knowledge and technology outcomes. We review what we know from our own experiences in systematic case studies, especially NSF-sponsored university research centers.

A second observation is more critical of the science of team science literature. Many of these findings of this literature can be misleading because they address best practices rather than general processes and activities that can be facilitated by a number of practices. For example, should face-to-face proximity be required of all instances of team science? In our view as organizational and science policy scholars the answer is "it depends," i.e., only in cases when that particular feature is required to facilitate one or more forms of the organizational capital that is *required* (Ouchi 1980) to facilitate collaborative research amongst diverse sets of human capital and other inputs. Should all instances of team science avoid formal structures? Again our response is the same, only to the extent to which there is no informal organizational capital, like congruent goals and resource interdependencies, already at play. Our basic point is that team science at the National Cancer Institute may need to look very different than team science at the National Institute on Alcoholism and Alcohol Abuse.

In contrast, the strategic management literature discussed above focuses much less on "best practices" but rather emphasizes a number of practices that can bring about one or more types of organizational capital for coordinating diverse actors. For example, participatory decision making may be facilitated by face-to-face meetings (which the science of team science literature emphasizes) but also in a number of other ways, including but not limited to leadership styles (Northouse 2009) and project planning activities (Lipnack and Stamps 1994). And there are times when formal structures and authorities are required for coordinated problem solving, including but not limited to cases of "radical" research and development (Ettlie et al. 1984; Damanpour 1996).

3.1.2   Strategic Management and Leadership at the Organizational Level

Before getting to the cooperative research centers literature, we address the strategic management literature on organization-level strategic management and leadership in research and development. The management science literature emphasizes the critical underlying organizational and behavioral mechanisms that the centers literature has for the most part overlooked but is starting to address more explicitly. Management science focuses predominantly on firms performing R&D as a core activity, and has done so for decades. Generally, the findings of studies of these firms demonstrate them to have relatively "organic" or "clan" structures characterized as decentralized, informal, but highly integrated by goal congruence and resource interdependence, i.e., the informal organizational capital addressed in the logic model in sect. 2.3 (Ouchi 1980; Laursen 2002; Laursen and Foss 2003; Cano and Cano 2006). The rationale underlying this way of organizing is that firms focused on R&D require collaboration among workers with expertise in different knowledge areas, which makes it difficult to predict (and therefore govern) individual contributions to production (Schumpeter 1934; Aiken and Hage 1971; Souitaris 1999; Laursen 2002) with formal structures and authorities for governing behaviors (Itoh 1993, 1994).

There have been few studies addressing "within group" differences for firms focused on research collaboration as a core activity. Ettlie et al. (1984) and Dewar and Dutton (1986) find "incremental" research (not deviating much from existing knowledge and practice) to correlate informal organizational capital and "radical" research (deviating greatly from existing knowledge and technology) to correlate with formal organizational capital. This latter relationship occurs because there is low goal congruence and resource interdependence amongst collaborators, which requires structure to keep individual behaviors aligned with firm goals due to the risk that is inherent in relatively "radical" departures from existing knowledge and practice (Ducheseneau 1979; Ouchi 1980; Hage 1980).

A subset of the strategic management literature focused on firm-firm collaborations, studies of human resources management (HRM) practices typically address the relationship between HRM practice and firm performance in R&D. Many studies emphasize a positive relationship between R&D performance and "strategic" HRM practices—typically ones including decentralized or participative decision making, team-based R&D, continuous learning, job rotation, training and development, and incentive-based approaches to R&D outputs (Laursen and Foss 2003;

Cano and Cano 2006). These practices are not interpreted as "best" in this literature, but rather as a number of options for generating the informal organizational capital emphasized in our logic model, e.g., congruent goals and resources interdependence. In contrast to these practices are "technical" HRM practices not emphasized in studies of firm performance in R&D, which include functions such as hiring, firing, promotion (Huselid et al. 1997; Lepak and Snell 1998; Lepak et al. 2005). Notably, centers lack these authorities, which may be considered "strategic" in the absence of goal congruence (Boardman 2012).

A growing proportion of the literature on cooperative research centers—typically those that are university based—are emphasizing effective (and ineffective) management and leadership at a fundamental level—i.e., in ways that can be brought about by a number of practices—rather than by emphasizing discrete practices as "best." Chompalov et al. (2002) identify four basic approaches for managing research collaborations and these types adhere to the different combinations of organizational capital—goal congruence, resource interdependence, and formal authorities—emphasized as fundamental for coordinating inputs and resources. "Bureaucratic" collaborations are most successful when the collaboration involves multiple organizations and there must be structural component to ensure that no single organization's interests are disproportionately served. Leaderless collaborations are too structured with rules and procedures for coordination and collaboration but without the hierarchy protecting against dominance by one organization or another. This type works when there are other informal mechanisms keeping collaborating organizations "in line" but yet some rules and procedures for coordination across organizational boundaries are required.

Since Chompalov et al. (2002), few studies have assessed variation in the management and leadership of diverse sets of actors across multiple cooperative research centers and contexts, i.e., from a logic model perspective, few general studies assess the "processes" of centers. Studies of center processes have focused predominantly on specifying management challenges rather than solutions. Management in centers has been described as ad hoc and underdeveloped (Bozeman and Boardman 2003). There is evidence of negative attitudes towards commercial research (Boardman and Ponomariov 2007) and shirking (Boardman and Bozeman 2007) among center faculty due incongruence between faculty goals and center goals. Goal congruence is a formidable management challenge in NSF centers in particular, as most faculty in these centers have primary appointments in academic departments (Bozeman and Boardman 2003). Toker and Gray (2007) demonstrate challenges associated with spatial dispersion in NSF centers. How centers are responding to these challenges—and how these responses relate to center outputs and impacts—goes generally unaddressed in the literature.

Boardman (2012) describes NSF university-industry research centers as both dysfunctional organizations and organizational networks. NSF centers share attributes characteristic of collaborative networks and formal organizations. Like networks, all NSF centers are boundary-spanning. Though NSF centers vary widely in terms of research foci (Stahler and Tash 1994) and policy goals (Bozeman and Boardman 2003), all are "problem focused," emphasizing not the extension of knowledge in a particular field or discipline per se, but rather the production of new knowledge and technology to address social and economic problems (Boardman and Gray 2010). This focus on problems rather than disciplines sees that NSF centers are also boundary-spanning, involving faculty from multiple disciplines (and thusly from multiple academic departments and universities) and from multiple sectors (including industry, universities, and government). Boundary-spanning collaboration is regarded as the primary justification for funding centers, at the NSF and elsewhere (Becker and Gordon 1966; Ikenberry and Friedman 1972). Like new organizations, NSF centers must create leadership and authority structures as a concomitant to setting up new organizational designs. All NSF centers programs publicly solicit for proposals from teams of university faculty. Each center grant that is awarded is university-based and to be lead by the principal investigator(s) who submitted the proposal. Proposals for all NSF centers programs are required to address internal organization and management, including horizontal and vertical differentiation and authority lines. However, leaders of NSF centers face a confluence of management constraints that can hinder coordinated problem solving as it occurs in both networks and formal organizations. A major gap in the literature is how NSF centers are using organizational capital to facilitate collaborative research and development amongst and within university-industry and multidiscipline projects.

Chapter 4
The Outputs, Outcomes and Impacts of Research Collaboration

4.1 Introduction

In this chapter we consider impacts in three inter-related categories, impacts on knowledge-focused products, impacts on property-focused products and impacts on scientific and technical human capital (or "capacity-building products"). With respect to the first two categories, we recognize that outputs from a great many research projects and collaborative efforts have both property and knowledge goals (see Hessels and Van Lente 2008). Nevertheless, this distinction retains analytical convenience, in part because it is reflected in much of the literature.

We later examine the growing literature on the negative impacts of collaboration. While many seem to assume that research collaboration is an inherently beneficial activity, it is the rare human enterprise that has no negative impacts whatsoever. For example, innovation is prized by most public policy-makers and business owners but we are quite familiar with the sometimes dual edged nature of innovation- creating new jobs, but eliminating the need for others, creating a cleaner and healthier environment, but sometimes degrading the environment, stimulating new energy sources and energy efficiency and helping create energy shortages and, sometimes, energy-related disasters. Much the same set of contrasts and paradoxes obtains for research collaboration outputs and impacts, but these are less studied and, among those who do study impacts of research collaboration the negative focus is most often on industry-university collaborations, particularly alleged problems created for faculty and students (e.g. Slaughter and Leslie, 1997; Slaughter, et al. 2002; Kleinman 2003). We later suggest some additional avenues of study for those interested in the negative impacts of research collaboration.

B. Bozeman, C. Boardman, *Research Collaboration and Team Science,* SpringerBriefs in Entrepreneurship and Innovation, DOI 10.1007/978-3-319-06468-0_4, © The Author 2014

4.2   Outputs and Impacts for "Knowledge-Focused" Collaborations

In most fields of inquiry, studies focusing on public policy, social or business enter-prise impacts, or measures of impact often disappoint owing to scarcity, undue com-plexity or indirect measurements and attendant assumptions, sometimes misplaced, about the ability of the available measure validly to substitute for the behavior or event one wishes ideally to measure. The study of research impacts and, by exten-sion, research collaboration impacts, fares well on this account. While a great many measures have been suggested in connection with research output or impact, two types of measures stand out as, despite problems we shall explore here, having some face validity and intuitive appeal. In the case of knowledge-focused collaborations the use of citations has become widespread and increasingly powerful. In the case of property-focused collaborations, patent statistics often prove valuable. The vast majority of empirical studies of research impacts focus on one or both of these types of measures. The result is an increasingly coherent "dependent variable" but, at the same time, some predictable gaps in research.

We begin by examining studies that focus on knowledge products pertaining to research collaboration. The worldwide revolution in the use of bibliometrics, now decades old in its development (Garfield 1979; Hood and Wilson 2001; Godin 2006), has been a great boon to the study of research impacts. Nevertheless, biblio-metrics-based research has some important limitations with respect to understand-ing outputs and impacts of research. The chief problem with bibliometrics-based studies is not what they capture but what they fail to capture. In most cases, biblio-metric studies do not even aspire to providing information on knowledge impacts. Many studies (e.g. Hummon and Dereian 1989; Onel et al. 2011) focus on the re-lationships of citations to social networks. Still other studies focus on citations as a direct and immediate dependent variable rather than as an expression of knowledge. Thus, for example, studies examine the effects of various dimensions of co-author-ing teams on the accumulation of citations (Katz and Hicks 1997; Persson 2010).

To the extent that leaps in knowledge track against citations, bibliometric studies are valid and revelatory. But there are known limits to what can be captured by cita-tions including, for example, the social and economic implications of indicators of "highly cited." Most agree that citations can prove an excellent starting point; work that is not cited does not necessarily warrant ignoring, but a lack of citations is an excellent indicator that it has, in fact, been ignored. Nor can we assume that works that has been highly cited have been widely used in any conventional meaning of "use of knowledge." There is some evidence that citations are often social signs (Cronin 1984) and that many works cited are not actually read by the person citing (though many commonly scholars who do not read entire articles do read abstracts (Whissell 1999)). The fact that citing authors are not intimately familiar with the work of the cited author does not mean that the latter's work has no influence, only that its influence may not be entirely substantive, that influence may be indirect, and that the influence of the knowledge itself (i.e. the embodied propositions, evidence and argument) may be confounded with any of a wide variety of social determinants of citation behavior, including personal relations, reputation of the author or the journal or even the page upon which a citation occurs in a Google Scholar search. This does not necessarily diminish either the authority or scientific standing of highly cited papers or of highly cited researchers; the point is simply that one should not draw an equation between citations and the growth of knowledge. Citations provide a good starting point, a measure that is precise, consensual, normatively meaningful, and, most important of all, easily available.

This cautionary preamble is merited because so many research evaluation and research collaboration studies use citations, often to good effect, as indicators of output or impact. A general overview of research collaboration impacts (e.g. Bozeman et al. 2013) would perhaps include scores of such studies. However, since the focus here is chiefly on industry research, industry-university collaborations and, generally, boundary-spanning collaborations, we review only a small portion of the citation/output/impact studies.

Most citation-based studies of research output or impact focus squarely on academic science. Few if any "what-does-industry-want-from-its-university-collaboration?" studies list citations as a "want." At least one study (Lee 2000) finds that industry's collaborations with universities are strongly motivated by a desire to benefit from the fundamental knowledge developed by universities, but this does not imply that citations counts or impacts of citations reflect the knowledge uses intended by industrial partners (Bozeman and Rogers 1997). Other studies (e.g. Feller and Roessner 1995; Feller et al. 2002) suggest the industry is more interested in access to individual scientists and to graduate students (sometimes with intent to hire them). If the intent is to directly appropriate research, the concern of industry partners often has as much to do with cost avoidance (Gray and Steenhuis 2003), equipment access (Tartari and Breschi 2011) or with the fit of the research to near-term needs (Bozeman and Wittmer 2001) as with the "best" or "cutting edge," at least as these are measured with citations.

To be sure, there are studies that focus directly on the research collaboration impacts on (citation-measured) productivity. However, almost all these studies examine university-affiliated co-authorships rather than collaborations between academic and industry scientists or collaborations among industry scientists only. There may be lessons to be gleaned but it is not clear these lessons relate directly to industry-only or industry-university collaborations. For example, Lee and Bozeman (2005) study of hundreds of academic scientists found, perhaps relevant) that the relationship of collaboration to citation-based measures of productivity is not at all straightforward. Collaboration has a strong and salutary effect on productivity if one is measuring productivity in terms of the number of articles produced (normal count) but if one reduces that number in proportion to the number of authors (fractional count) then the seeming productivity gains of collaboration vanish. The study also found that younger and mid-career scientists have greater productivity pay off from collaboration and that collaboration affects job satisfaction. But it is not possible to determine if these findings transcend their context, chiefly collaborations among academic scientists.

Another study (Pravdić and Oluić-Vuković1986) *potentially* relevant, but limited by its academic science context, provides citation-based evidence that it is not only collaboration that has an impact on scientific output (in terms of the number of scientific articles produced), but also the tendency of authors to maintain a more stable set of co-authors and collaborators. While the authors do not provide hypotheses about possible causes, one can envision alternative explanations. It is possible that co-authors develop increasing trust or an improved ability to work together or that teams simply sustain diminishing transactions costs for collaboration, thereby making collaborations more efficient. It is also possible that teams stay together to wring out the "least publishable unit" (Susser and Yankauer 1993) publication from their project, therefore elevating productivity only on the "more articles published" basis.

Despite these various caveats, there are several studies of knowledge-focused collaborations that provide more encouraging results. Caloghirou et al. (2001) study of university-industry collaborations in the European Framework Programs showed that firms reported substantial enhancements to their respective knowledge bases, indeed this was listed as a first benefit, followed by improvements to production processes. The study examined a large set of research joint ventures occurring during a fourteen-year period. They found that firms from smaller economy European nations were more likely to participate and more likely to benefit. Benefits occurred not only in knowledge acquisition, but also cost savings, cost sharing, and cost avoidance. But the chief benefit reported was increments to the firms' knowledge base. The study was based on firm reports with no other supporting data.

Levy et al. (2009), a study sharply limited by its focus on industry collaborations in a single university, provide a finding quite relevant and, if it has external validity, important to the study of industry-university research collaboration. According to their analysis there is a strong tendency of industry-university collaborations, whatever the initial intent, to center on knowledge-focused output rather than product-focused output. In a different European setting, and with a great deal more data, Ambos (2008) and colleagues arrive at similar conclusions about the tendency of academics to push industry collaborative work away from property-focused to knowledge-focused research. In light of the incentive systems (Fulton and Trow 1974; Ponomariov and Boardman 2010) within most universities, not to mention the tendency of researchers to stay in their "comfort zone," it is perhaps not surprising that such collaborations would tend to drift toward knowledge-focused (read: "publications-focused") outputs.

4.3 Outputs and Impacts for "Product-Focused" Collaborations

Collaborations centering on developing intellectual property and, ultimately, commercially viable products and services, include a wide array of organizational and institutional mechanisms, ranging (at the low end of formalization) from the informal collaboration of researchers (Link et al. 2007) in different sectors to such mechanisms as joint ventures, research consortia, multi-discipline/multi-institution university research centers (MMURCs), formal research alliances and partnerships, co-location incubator cooperative research, and research and technical assistance contracts. This list is not exhaustive (for an expanded list see Vonortas 1997; Hagedoorn et al. 2000; Hoekman et al. 2010). Not surprisingly, given the importance of collaboration to commercialization objectives, there is a considerable relevant literature about the benefits of collaboration. However, many such publications are reports of personal experiences, unsystematic case studies or, most common, conceptual models based on no data. While there are many evidence-based studies as well, the literature is somewhat fragmented and varies a great deal in its indicators of outputs and impacts. Aside from data based on personal reports, the most common evidence marshaled in the studies of property-focused collaboration is patent data, including patent citations (van Zeebroeck and Potterie 2008).

In considering benefits from university-industry collaboration, it is useful to establish a baseline, namely the productivity rate when *not* collaborating. Morgan et al. (2001) provide such baseline information, albeit by this time somewhat dated. Using National Science Foundation data from the 1995 Survey of Doctoral Recipients and the 1995 National Survey of College Graduates, the authors analyzes responses from 204,700 scientists and engineers working in government or industry. While a variety of technical activities are examined, the chief focus for present purposes is patent rates.

Morgan and associates report that the patent activity rate (filed, but not necessarily awarded) for doctoral level scientists and engineers doctorate holders in all employment sectors is 12.1%. In industry, 20.8% were named as inventors on patent applications during the five-year period since April, 1990, more than triple the rate (6.0%) of those those working in universities. Importantly, while almost all those filing patent applications from academia are doctorate-holders, industry 44.6% have bachelors degrees, 26.2% master's degrees and 29.2% have doctorates. Their analysis shows that industry-based scientists and engineers have a higher rate of success developing products from patents, with 30% of patents yielding (within five years) a commercial product, process or license, compared with the academic scientists and engineers where about 20% of patents awarded resulted in a commercial product. However, those working in higher education has a slightly hire patent granting success rate than those in industry. In academia, 41% of those named as inventors in at least one patent application in the five years studied received at least one patent, compared to a 36% "hit rate" among industry scientists. Given these baseline (no-collaboration) data, how does university-industry collaboration patent success rates compare? No truly comparable data are available.

There is little contention as to whether academic research and collaboration with industry feeds industrial organizations' stock of intellectual property (Mansfield 1999), even when there is not an active collaboration between academic scientists and industry. But there is also widespread agreement that the development of research-to-product can in many instances be improved and accelerated with active collaboration, including patent development and utilization (Henderson et al.

1998) but also collaborations not based on patents but on such goals as research and technical assistance, public domain research, research based on trade secrets and, often over-looked, collaborations on manufacturing technology and other production processes.

If we assume, as seems warranted (Audretsch et al. 2002), that university-industry research collaboration often (though not inevitably) provides important benefits, then a next question is who benefits and under what circumstances? Lööf and Broström (2008) find that large manufacturing firms benefit most from collaborating with research, whereas smaller firms and service firms tend to benefit modestly if at all. This finding is consistent with a series of studies by Bozeman and colleagues, but ones focused on the benefits to firms working with federal laboratories on cooperative research agreements (see Bozeman and Papadakis 1995; Rogers and Bozeman 1997; Crow and Bozeman 1998; Bozeman and Wittmer 2001; Saavedra and Bozeman 2004). These studies examined patent and license creation, firm participants' reported monetary costs and benefits, and experiences developing new products. By most indicators, first benefiting most were ones that were larger, were themselves active in research collaboration (rather than simply using federal laboratory results) and, interestingly, ones who did not have as a primary goal the near term appropriation of intellectual property. Small first tended to be much more focused on developing products from the cooperative research and, while this happened more than for larger firms, the small firms also reported lower satisfaction and lower ratio of benefits to cost.

One old (pre-Bayh-Dole legislation) but still widely used mechanism of industry-university collaboration is industry funding of university R&D. The impacts of such funding remain unclear because relatively few studies have been performed. One recent study (Hottenrott and Thorwarth 2011) of German companies' funding of university research indicates that there is a "skewing problem" (the authors' terms), namely that university researchers do alter their intended research trajectories in the service of industry contracts' objectives, arguable to determine effects on the development of knowledge-focused research. However, the authors also find that industry funding has a positive effect on applied research, at least as measured by patent filings and patent citations. Moreover, another study, using a somewhat different database, found no evidence of a skewing problems and at least some evidence that industry funding has a positive effect on knowledge-focused research.

Most university-industry collaborations are not centered on publications and do not yield patents. Even among university researchers who are strongly oriented to working with industry, most do not work with industry in developing patents. Bozeman and Gaughan's (2007) data from the Survey of Academic Scientists, shows that among "industrially-active" academic researchers only about 5% have co-developed a patent. Lissoni and colleagues, focusing on three European countries, report that a little more than 4% of those working with industry have developed a patent from their collaboration. In both studies, results vary considerably by field and by the nature of the university setting (Kenney and Goe 2004) or its organizational culture (Agrawal and Henderson 2002), but patenting is always a relatively rare outcome from university-industry outcomes.

In the U.S. much more common activities include collaboration of research, consulting, technical assistance, and placement of students in industry (see Lin and Bozeman 2006). According to Bozeman and Gaughan's findings for a national sample of academic STEM faculty in Carnegie-Extensive universities, approximately 17% had obtained an industry grant or contract in the 12 month period preceding the study, just about the same percent (18%) who had engaged in industry research consulting during the same period. Studying German academic researchers Grimpe and Fier 2010) report quite similar results with about 17% working as paid consultants, a similar amount reported in Haeussler and Coyvas' (2011) study of United Kingdom and German university scientists working in the life sciences. A study (Klofsten and Jones-Evans 2000) of science, engineering and medicine faculty in Sweden and Ireland reported a high rate of industry interaction, but most of these interactions (68%) these being limited to consulting and information exchange.[1]

4.4   Outputs and Impacts for Scientific and Technical Human Capital Impacts

Why should policy-makers be concerned with the STHC impacts of collaboration independent of the effects of collaboration on the production of knowledge and property? STHC is about the ability to *sustain* benefits from production of knowledge and technology and, thus, is a particularly useful perspective from a policy-making and planning standpoint. Moreover, research collaboration, including boundary-spanning collaboration, has strong potential for affecting STHC, in some cases positively in other cases less so.

Since there is relatively little research proceeding directly from a STHC perspective, the research directly relevant to research collaboration is quite modest, but worth examining. At least a few studies (Bozeman and Corley 2004; Gaughn and Robin 2004; Corolleur 2004; Davenport 2004; Dietz and Bozeman 2005; Lin and Bozeman 2006; Woolley et al. 2008; Boardman 2009; Ponomariov and Boardman 2010; have specifically taken a STHC perspective on scientific outputs and career development and several of these studies examine research collaboration directly or they are relevant to collaboration issues.

To succinctly reiterate, the STHC model focuses on the conjoining of researchers' S&T-related human capital (e.g. formal training, experiential knowledge, skills) and their S&T-related network ties and resources and relevant social knowledge (e.g. assessments of others' research foci, interests and abilities, knowledge of the "market" for research products, knowledge and access to particular channels for distributing knowledge). It is assumed that a researchers' productive capacity is reflected in accumulated STHC and, further, that STHC changes regularly and in some cases even predictably with, for example, exposure to additional knowledge, additions or subtractions from social networks. It is further assumed that STHC

---

[1] Some of the results reported in this paragraph are draw

reflects the vast majority of researchers' capacity to produce knowledge, with the other important factors relating chiefly to resources (e.g. grants, salary, research assistance) and organizational arrangements (most important, a job and a means of making a living from research but also particular attributes and incentive systems provided by institutions). If one knows these things then, arguably, one knows all one needs about individual productive capacity. Moreover, STHC and concomitant productive capacity can be viewed as additive or at least multi-level. Thus, we can refer to an individual's STCH but also to the accumulated STHC of a work group, a lab, a firm or a university research center. The only qualification, not a small one, is that accumulated STHC is mediated by the ability of the "human pieces of the organization" to not only fit together but to share and build on one another's STHC. In sum, STHC provides a useful perspective for understanding research collaboration; to some extent at least, collaboration is a vital ingredient in organizations' ability effectively to deploy the STHC of those employed by or affiliated with the organization (for amplification see Dietz et al. 2001).

One early study (Dietz and Bozeman 2005) of the relation of STHC to collaboration focused specifically on the value added of experience in industry to productivity and collaborations. The authors analyze career mobility, specifically movement between academic careers and industry careers but also medical careers and government positions, to determine possible productivity effects. They use two data sets, first data from the curriculum vitae of 1,200 research scientists and engineers, which they combine with patent data from the U.S. Patent and Trademark Office. They classify job transitions according to the sector of origin and the destination sector. For this set of scientists and engineers, the most common job transition is academic-to-academic (62.5 %) followed by industry-to-academic (8.2 %). The authors find that both inter and intrasectoral changes in jobs throughout the career result in higher STHC that, in turn, yields higher research productivity measured in terms of publication and patent rates and impacts.

Several papers on scientists' and engineers' professional networks have bearing on the core questions of STHC. A literature review by Audretsch et al. (2002) concludes that research evidences shows that firms with network ties to universities (apart from formal collaborations) tend to have greater R&D productivity as well as a higher level of patenting and that these relationships provide greater capacity to the companies but also to academic faculty and students. In another relevant study of STHC, networks and productivity, Goel and Grimpe (2011) examine both "active" and "passive" networking, wither active networking measured in terms of participation in research conferences. While this presents problems inasmuch as different participants have diverse experiences at research conferences, it does provide a useful contrast to their measures of "passive" networking that pertains chiefly to interactions with ones geographically proximate colleagues. Their findings indicate that passive networking is both complement and substitutes for active networking depending on the particular character of passive networking and the setting in which it occurs. In particular, research group leadership, a passive form in their terminology, is complementary. The results indicate that not all forms of network behavior contribute substantially to STHC.

Martinelli et al. (2008) emphasize the importance of external relations to collaboration, arguing that academic researchers who have few external ties (i.e. S&T human capital) have especial difficulties developing collaborations. Another study (Nilsson et al. 2010) focuses on the role of universities in providing supportive infrastructure for building STHC, arguing that the university incentives and environment determine a great deal about the academic researcher's likelihood of developing STHC as part of industrial networks and activity. Related et al. (2008) find that informal interactions between university scientists and private sector companies trigger more formal and more intense collaborations with industry. This reinforces earlier findings from Link, Siegel and Bozeman about the importance of informal university technology transfer in collaborations with the private sector.

Ponomariov and Boardman (2010) provide information directly relevant to the formation of STHC as related to research collaboration. The authors find that institutions have strong effects on STHC and collaboration patterns. Focusing on multi-institutional university research centers they find that such centers are effective not only in enhancing overall productivity but also collaborations and these cross-sector collaborations are especially likely to yield growth in STHC of collaboration participants. In a more recent study, Bozeman and Boardman (2013) find that such university research centers serve as an important nexus for collaboration and that they serve to change STHC not only by introducing students to a wider array of work possibilities but also by providing additional support for a wider array of learning experiences for faculty, students and center affiliates.

4.5 Negative Impacts of Boundary-Spanning Research Collaboration

Much of the controversy about possible negative effects of industry-university research collaboration centers not on research but universities' teaching missions. Perspectives diverge sharply. By one view, chiefly based in the management and economics literature, university interaction with industry is viewed as at worst inefficacious (e.g. not yielding intellectual property) but never harmful, expect perhaps in the most extraordinary instances. To be sure, the studies examining cooperative research, patenting and technology transfer almost never directly examine possible side effects or "collateral damage," especially in the teaching realm.

4.5.1 Possible Negative Effects on Teaching and Education

One finds an entirely different perspective from the higher education and sociology of science literatures, with many works providing sweeping critiques of university-industry collaboration, may focused on teaching and mission drift. The most familiar critique of "academic capitalism" is offered Slaughter and Rhoades (1996) and

Slaughter and Leslie (1997), who pull no punches in their arguments that industry-university collaboration, including ones based in university-based cooperative research centers often detract from traditional educational missions and goals. This occurs, according to the argument, as market driven-faculty increasingly shirk both traditional research duties and, especially, their teaching duties in search of profit or currying favor with industry (Slaugher et al. 2002). At least some of the academic capitalism critique is based on evidence, generally interviews or case studies.

To be sure, the arguments of academic capitalism critics bear examination and it is likely there are instances where either teaching or research is impaired due to work with industry (or government). But at least one body of evidence (there is scant research literature on this topic), coming from the Research Value Mapping database and the Survey of Academic Scientists, suggests that negative effects on teaching outcomes may not be commonplace. Lin and Bozeman (2006) show that academic researchers who are "industrially active" tend to provide more research support and assistantships to students and to be more likely to place them in industry jobs. A more comprehensive study (Bozeman and Boardman) of effects of research center-based collaborations, including those with industry, shows that industrial involvement appears to have a net positive effect on education Bozeman and Boardman (2013). These findings show that faculty who are more involved in industry collaborations tend to support more students, both graduate and undergraduate, place more students, and have a stronger mentoring orientation. Surprisingly, those who are industrially-active have no decline in the number of students taught and are likely to advise more students than others. Those with government grants, by contrast, are somewhat likely to teach less and to teach fewer students (Bozeman and Gaughan 2007). To some extent these perhaps unexpected findings are owing to the concern of university executives and public policy-makers that university research centers do not undermine teaching missions. For example, NSF centers programs often include requirements for demonstrating positive effects of center activities on student learning, training and career support (Ponomariov and Boardman 2010).

4.5.2 Possible Negative Effects on Research and Research Careers

The literature on possible negative effects of university-industry collaboration on research and research careers is much more extensive than that focused on effects on teaching and education. One common theme, with consistent research results, is that collaboration tends to deflect academic scientists from the research they would otherwise be doing and this, typically, is a substitution of more applied or development-oriented work for more basic work focused on knowledge development and less so on research or technical activity aimed at developing new products or services. There are two obvious problems with the research and this line of reasoning. First, not everyone agrees that fundamental research is more inherently or inexorably more valuable than applied or product-oriented work and, related, it is likely that the issue is not that one is more important than the other but that they should be provided in some optimal mix. However, no one has yet determined (or is likely to do so) that "optimal mix," though it is an issue that is much discussed (e.g. Reagan 1967; Amick 1973; Link and Long 1981; Nelson 1982; Banal-Estañol and Macho-Stadler 2010). Second, *any* research focused on an inherently unverifiable counter factual has limits to its validity (Chellas 1975). Moreover, even given these limitations, there is little evidence of major changes in research foci. Hicks and Hamilton (1999) in their study of the composition of university research after the passage of the Bayh-Dole Act, found no diminution in basic research publications (see also Hicks et al 2001).

Regardless of the mix or perceived social value of various types of work, there is the issue of whether academic scientists collaborations with industry suppress publication. The evidence is for the most part that it does not. Faculty with industry support tend to publish as many or more articles as their peers without industry support (Blumenthal et al. 1996; Gulbrandsen and Smeby 2005) and, according to some, such collaboration often yields new ideas an approaches as both industry and academic scientists bring in fresh perspectives of persons with different backgrounds, interests and motivations (Perkman and Walsh 2009).

One study (Azoulay et al. 2009) is dismissive of the idea that industrial involvement and particularly patenting affects negatively the research work of academic scientists. Focusing on university patenting, but not inter-sector collaboration, employs data from 3,862 academic life sciences finds that patenting activity is positively associated with publication of scientific articles and positively, but weakly, associated with the quality of those articles. The vast majority of empirical research (e.g. Agrawal and Henderson 2002; Azoulay et al. 2007; Breschi et al. 2007; Fabrizon and Di Minin 2008) supports the idea that academic scientists industrial activity has no suppressing effect on productivity, variously measured, and may enhance it.

Regardless of the level of research activity, one must also consider whether academic collaboration with industry does, in fact, change the research agenda. While it is not easy to sort out this issue, there is some considerable evidence (Blumenthal et al. 1996) that work with industry does change research agendas, usually, as one might expect, toward industry needs and usually in the direction of more applied research. However, there are also studies (e.g. Van Looy et al. 2004) that show no significant alteration of research agenda due to industry support. The difference may relate to distinct characteristics of scientific fields and perhaps to national context.

In some cases there is a potential for suppression or late publication of research (Blumenthal et al. 1997). There is substantial evidence of this occurring; with relatively modest effects in most STEM fields (Murray and Sterns 2007; Huang and Murray 2009) and much more pronounced effects in biotechnology and medical research (e.g. Campbell et al. 2002; Lo and Field 2009). Bekelman et al. (2003) provide a meta-analysis of biomedical literature on possible conflicts of interests and their effects.

Just as great a concern is whether the commercial activity in academic science reduces the growth rate of public domain science. Using data from the U.S. Small Business Innovation program and control groups for randomly selected papers during the period 1994–2004, Toole and Czarnitzki (2010) find that there is a "brain drain" away form public domain research and conclude that this could in the long run have damaging effects not only on the amount and availability of basic research that it could have damaging effects for economic growth if the proportion of public domain knowledge continues to shrink.

A body of work investigates the idea that collaboration results in industry siphoning off STEM research talent. The concern is that scientists who would otherwise focus on public domain science move to industry careers and, related, that highly trained industrial scientists are especially likely to feel alienated and dissatisfied with their jobs (Miller 1967). Recent work (e.g. Roach and Sauermann 2010; Fuhrmann 2011; Agerwal and Ohyama 2013; has provides a richer understanding of academic and industrial research careers. In the first place, the idea of job choice is itself distorted in some analyses that fail to recognize the role of demands and structural barriers to both academic and industrial jobs. Recent research by Sauerman and Stephan (2013) questions the "conflicting logics" view of industrial vs. academic science careers. Their study more than 5,000 researchers in industry and universities shows that there is nearly as much normative and structural heterogeneity within the respective sectors as between them and that the sector context does not explain as much about the nature of research and research careers as has previously been assumed.

For many years, the research management literature has assumed that scientists moving from academia to industry make a key choice, sacrificing autonomy and gaining material rewards, including higher salaries. In their recent study Sauermann and Roach (in press) examine this conundrum and provide some evidence based on data from nearly 2,000 STEM doctoral students poised to enter the job market. They develop a model of the "price" of foregoing published research and find that it is closely related to job aspirants' views about their own research abilities, but also desire for peer recognition, desire to contribute to society and, of course, their preferences for money and higher salaries. Even in industrial jobs where publishing is encouraged, those who have these characteristics and who are from leading institutions are more expensive to hire.

Another recent career choice study (Agrawal and Ohyama 2013) uses life cycle models of human capital to understand the sorting of researchers into academic or industry careers. Using Data from the National Science Foundation's Scientists and Engineers Statistical Data System they find that university scientists sorting between basic and applied science is based on ability (human capital) but ability does not explain such sorting among industry scientists. Likewise, the earnings trajectories for basic and applied scientists differ in academia but are quite similar in industry. In general, the study corroborates the widely held assumption that academic scientists tend to have lesser material wealth motivations than those choosing industry careers. However, it is perhaps useful to note that sorting into jobs is not merely a function of values and preferences. Fox and Stephan (2001), in their study of 3,800 doctoral students in STEM fields, analyze career preferences and outcome patterns according to gender and field and fine that these strongly affect expectations about science career possibilities. Other studies about choices between academic and industrial research careers show that there are strong peer and collaborator effects. Stuart and Ding (2006) analyze careers of life scientists and show that one of the most important predictors of their own preferences for work with industry and for entrepreneurial values is very much a function of the social-intellectual context; if peers and mentors embrace entrepreneurial values, then the individual is likely to do so. This implies, for good or ill, depending on ones values, that as universities become more focused on industrial interaction and entrepreneurial work then younger STEM researchers, including graduate students and postdoctoral researchers, are more likely to embrace these diffused values.

4.5.3   Negative Impacts on Industry

The literature examining possible negative effects of boundary-spanning research collaborations on industry is quite small, with most studies relevant to this topic not taking it as a chief objective of the research. It stands to reason, of course, that industrial partners could often suffer negative impacts of a variety of sorts, including loss of funds, opportunity costs, limited productivity investments and even infringement on proprietary knowledge and technology.

One possible negative impact on industry is nonproductive uses of time as industrial and university partners dispute intellectual property ownership (Siegel et al. 2003) or more often take inordinate amounts of time to clarify and agree upon ownership terms and conditions (Hertzfeld et al. 2006). These concerns are cited by industry officials in interview and questionnaire-based studies (Cohen et al. 1998; Schmoch 1999), but little systematic research has document the extent or even the veracity of the perceived problems. Moreover, it is at least possible that the firms' dissatisfaction with time delays and intellectual property difficulties is owing to beneficial protections or accountability needs of their partners. Likewise, when university partners complain of firms' efforts to control intellectual property and channel research work in particular directions, this may simply be a need to adjust the balance wheel of collaborations so that diverse partners with diverse objectives can benefit from the collaboration. There is a dearth of systematic research not only on the problems firms face in inter-firm and inter-sector collaboration but also little attention to the ways in which the firms evaluate their experience. Perkmann et al. (2011) provide a suggested performance measurement system for such interactions but one that seems not to have been applied by firms.

Chapter 5
Effectiveness Questions and Research Recommendations

5.1 Pondering the Effectiveness Questions

In Chap. 1 we identified a number of specific research questions and our overview is in part aimed at providing some preliminary answers to these questions. Here we revisit the questions and we then provide recommendations about research needed for the advance of research and theory about research collaboration and team science.

In each of the questions below we provide a broad assessment of the level of evidence for the question, summarize the preponderance of the evidence and then provide recommendations. In our study as in any study examining evidence and then making recommendations the data do not speak for themselves. We use our experience as researchers and in assessing literature to arrive at recommendations that we feel are consistent with the evidence provided here.

5.1.1 Research Management Approaches

To reiterate, the question here is: *What does the available research on university-industry research partnerships and within-industry team science tell us about effective research management approaches and partnership models that support positive team processes and successful scientific outcomes?* We frame this question in terms of broader strategy and policy deliberations. In doing so, we find that the evidence base is relatively modest. With some exceptions we have not focused here on the micro-level management issues of teams and groups (for a review of this research see Mathieu et al. 2008). Little of this extensive research pertains to boundary-spanning collaborations between firms or between universities and firms). Instead we focus on "effective research management" in interorganizational contexts. With this much more limited focus we draw two ready conclusions. First, the available evidence is minimal. Second, not only is there little theory or research on managing boundary-spanning, inter-organizational collaborations, such literature as does exists suggests that managerial practices in such contexts is often poorly thought out

B. Bozeman, C. Boardman, *Research Collaboration and Team Science,* SpringerBriefs in Entrepreneurship and Innovation, DOI 10.1007/978-3-319-06468-0_5, © The Author 2014

and haphazard. Possibly, but not necessarily, the first problem (research scarcity) affects the second (haphazard managerial practice and managerial structures).

There are a number of factors at the organizational and team levels associated with entry and performance in interfirm R&D alliances and industry consortia. As shown by Joskow's (1985) research, monitoring transactions is vital to success of collaborations, especially careful monitoring of terms of contracts. Lax contract management tends to be associated with ineffective collaborations. Given that contractual relationships seem to be becoming more common (Panico 2011), this "lesson" is perhaps more important than ever.

One of the most common findings pertaining to collaboration success, at *every* level and for nearly every type of collaborative institution, is that the parties to the collaboration must have a high degree of trust (Sonnenwald 2007; e.g., Creamer 2004; Hara et al. 2003; Sonnenwald 2003; McLaughlin and Sonnenwald 2005; Knorr-Cetina 1999; Krige 1993) and trust is vital for effective management of university-industry and within-industry teams and collaborators (Dodgson 1993; Zajac and Olsen 1993; Fulati 1995; Davenport et al. 1998; Tartari et al. 2012).

One important focus in recent years has been on the ways firms learn in inter-firm partnerships. Sampson (2005) suggests that firm learning is an output of inter-firm R&D alliances (firm-firm, not industry consortia) that enhances firm level STHC specifically in terms of ability to manage uncertainty and ambiguity. She calls this learning process "alliance management skills." Fontanta et al. (2006) examine learning from a different perspective, emphasizing proactive "searching, screening, and signaling" of the external environment by firms for opportunities to engage in research collaborations with public research organizations. This search process, conducted effectively, enhances the effectiveness of collaborations.

The motives underpinning firms' research collaboration have been examined by a number of authors. According to Sakakibara (1997) cost sharing and skill sharing motives are two distinct and pervasive motives for inter-firm research collaboration. Hagedoorn (1993) find similar results for firms joining R&D consortia for complementary knowledge exchanges and more recent research (Hayton et al. 2013) supports the dual motivation of costs savings and skill sharing. Prahalad and Hamel (1990) suggest that firms considering joining consortia benefit from other firms' portfolios of core technical competencies that can enhance their own technical competencies. Hamel (1991) more specifically suggests that firms benefit from consortia to internalize the technical skills and competencies of others.

Several studies have not only focused on geographically dispersed collaboration but have examined the impact of different degrees of dispersion (Knoben and Oerlemans 2006) provide a general theoretical overview on geographic dispersion and collaboration effectiveness, including but not limited to research collaboration). Hoegl and Proserpio (2004), in their study of 145 software development teams find that proximity is positively and significantly related to teamwork quality but varies across dimensions of quality. According to D'Este et al. (2013), it is not simple proximity that matters to research collaboration success but rather patterns of clustering, technological complementarity among firms and the interaction of clustering with complementarity patterns.

5.1.2 Learning from Failure

The question examined here is this: *What is known about the reasons for failure in university-industry partnerships and within-industry science teams?* Research focusing directly on failures of boundary-spanning collaboration is quite scarce. One could perhaps consider this question by assuming that factors determining failure are essentially the mirror image of those affecting success. While there is some merit to such an approach it is not idea. It does not take into account the possibility of threshold effects (X quantity of an attribute has no effect, $X+1$ has extremely positive effects) and it does not take into account interaction among variables related to effectiveness. Thus, we return to these issue below- topics needed more attention in the research literature.

Generally, the literature examining failure studies (Barkema et al. 1997; Beamish 1985; Dussauge et al. 2000; Franko 1971; Gomes-Casseres 1987; Killing 1983; Kogut 1989; Li 1995; Park and Russo 1996; Pennings et al. 1994) emphasizes the lack of formal authorities and structures (e.g., ease of entry and exit, difficult to monitor and enforce contracts) and also the lack of informal mechanisms (goal congruence, resource interdependencies *after entry)*. Some studies suggest that formal research alliances and consortia are inherently unstable when compared to intra-firm or less complex inter-firm collaborations. When inter-organizational collaborations rely on either formal or informal governance mechanisms rather than relying on both, failure is more likely. Oxley et al. (2004) and Li et al. (2011) identify "knowledge leakages" to competitors as a particular type of failure in those collaborations where firms share not only precompetitive but also competitive knowledge. Usually competitive knowledge is shared inadvertently but is nonetheless damaging. Collaborations facing such difficulties sustain increased transaction/coordination costs and monitoring costs and more often experience free riding and other collaborative dysfunctions (Garcia-Canal et al. 2003; Gong et al. 2007; Hennart and Zeng 2002; Hackman 1987; Steiner 1972). It is difficult, of course, to manage the delicate balance among knowledge security, trust and collaboration. Another recipe for failure is when firms do not screen their external environments for opportunities to collaborate with other units (including but not limited to firms) or signal the wrong sorts of motivations and goals to potential collaborating firms.

5.1.3 Intellectual Property

We consider here: *How do intellectual property and conflict of interest concerns affect the collaborative processes and scientific and translational outcomes?* Despite the rapid growth of studies of patenting and licensing, there is only a modest *empirical* literature relevant the resolution of intellectual property and conflict of interest problems. That is not to say that the topic has not been addressed. Most studies focused on resolution of IP disputes are in law journals (e.g. Blackman and McNeill 1997; Hovenkamp et al. 2002; Mandel 2011). Since these studies are not empirical they are beyond our purview.

Even among those few potentially relevant empirical studies, most studies concerned with intellectual property problems and their remediation have only limited empirical support for their prescriptions. Still, they at least frame important issues. Sampson (2005) examines problems in firm-to-firm IPR negotiations and contracts and suggests that a key to avoiding problems is "alliance management skills." She notes that even in firms that have quality management and a history of effectively managing their own IPR, new challenges of alliances require new and perhaps different IPR management skills. Interestingly, some researchers indicate that governments can play an important role in lessening the likelihood of collaborative failures due to IPR problems. Research by Narula and Dunning (1998) shows that government agencies sometimes are quite helpful in mediating IPR disputes among firms that are parties to R&D consortia.

5.2   Recommendations for New Research

While the multidisciplinary literature on team science and boundary-spanning research collaborations is relatively new, it is not at this point a small literature. Having attracted persons from a wide variety of disciplines, but especially economics, management and sociology, and having spawned a variety of journals focused on these topics, the literature has flourished. But as it has flourished it has suffered the same problems as nearly all bodies of work in the social sciences, namely, some topics are abundantly researched and others, even fundamentally important ones, are neglected. It is easy enough to understand how this has happened in the literature of team science and boundary-spanning research collaborations, because the driving forces for over-concentration and, at the same time, neglect are nearly identical to those affecting other fields of study. Specifically, scholars choose topics because of data convenience and availability, because topics become fashionable (Crane 1969; Bort and Kieser 2011) and, at least it is perceived, more likely to be accepted for publication (Martin and Diego 2009; Besancenot et al. 2011), because funding becomes available for certain topics, and, generally, research follows the law of low hanging fruit.

At this point, much of the low hanging fruit has been plucked and, in some cases, the harvest has been bountiful. If one wishes to know about academic scientists' collaborations, both one-on-one and with institutions, then much literature is available. Similarly, there is a considerable and growing research on industry relations with university research centers. Our knowledge of co-authorship patterns is considerable, having been abetted by the growth of bibliometric data and techniques and, likewise, we know a great deal about all topics that are knowable through patent and licensing counts and citations. But much work remains.

5.2.1 Recommendation 1: Meta-Choice

There is research about strategic choices made by parties to boundary-spanning research and technology collaborations and this research, while not plentiful, is generally quite useful. However, there is almost no research that tells us how participants choose among available collaborative institutions and modalities. Thus, we know something about why, in general, industrial firms decide to participate in MMURC's, but we do not know much about why they choose particular MMURC's or, even more important, why they choose to participate in the MMURC as opposed to an inter-firm joint venture, a research alliance or a contractual relationship with another firm. This is particularly unfortunate inasmuch as this type of question is "knowable." As we suggested above, self-reports are not helpful for interpreting over-determined events having multiple agents, none of whom has the "big picture." But usually the number of people who make decisions about whether or not to participate in a collaborative institution, and which particular one, are few in number and can, if the wish to do so, provide a good account of their reasoning and discrete actions.

The problem with the research gap is not a shortfall of methods, money or knowledge, but of access. If mechanisms can be devised to enhance access to the relatively few high-level executives who control these meta-strategies, then a great deal can be learned about collaboration that we do not now know.

5.2.2 Recommendation 2: Research on Institutional Failures and the "Dark Side" of Collaboration

We noted above that the literature on the collaboration of large, complex boundary-spanning collaborations is at best scant. The literature on the dark side of large-scale collaboration is non-existent. There has been at least some progress in the last decade regarding the dark side of collaboration when the collaboration focus is small-scale, especially co-authoring teams. One reason for this progress is a widespread recognition of emerging problems, especially conflict of interest and problems related to co-authorship. Many of the dark side studies are in the biomedical sciences (e.g. Rennie et al. 2000; Wainright et al. 2006; Cohen et al. 2004), where ethical issues have emerged around such issues as "phantom authors" and industry payments (sometimes pay-offs) for research.

Within the context of large-scale boundary-spanning collaborations, there are at least a few studies of the effects of conflict of interest, usually considering whether university researchers have been "corrupted" by their relations with industry. So far, the more systematic and empirical the focus, the less likely one is to find problems. Nevertheless, the dark side has in some cases been well documented for smaller collaborations.

The dark side issues with large-scale boundary-spanning collaborations go well beyond what is conventionally addressed in the literature. The literature tends to

focus on bad behaviors or bad outcomes for academic participants in collaborations (deflected research agendas, diminished educational role) while neglecting
the dark side for industry participants. We know from case studies that IP disputes
are sometimes fierce and damaging, but we have little knowledge from systematic
studies employing aggregate data. Similarly, we know that companies sometimes
suffer (and sometimes perpetrate) industrial espionage, but we have no systematic
evidence about the extent to which and ways in which this occurs in boundary-
spanning or inter-firm collaborations. It is not easy to study bad behavior and disas-
trous outcomes, especially for firms, but it would be useful to make a start. Research
might prove quite beneficial. If we again take the biomedical research collaboration
research that has focused on academics, the research has led to a number of reforms,
especially in journal editor's requirements but also professional associations and
even funding agencies (see for example Rennie 2000; Marusic et al. 2004; Pichini
et al. 2005; Devine et al. 2005).

5.2.3 Recommendation 3: Scientific and Technical Human Capital

We do not surprise with our recommendation that it would be useful to have more
work and more careful work pertaining the STHC aspects of boundary-spanning research collaborations. Studies of STHC collaboration, while not common, are available. However, the conceptualization has thus far been applied *empirically* only
in university settings. Studies focus on the relation of STHC to the development
of students, academic researchers' productivity (Lee and Bozeman 2004; Carayol
and Matt 2004), and career trajectories (Gaughan and Robin 2004). The one study
clearly relevant to business and STHC is Lin and Bozeman's (2007) study of the effects of industry experience on academic scientists' collaboration and productivity.

If we consider that research tells us that what most firms want from their inter-
sector, boundary-spanning collaborations is access to the knowledge and skills of
persons outside the firm, then clearly the STHC focus is quite compatible with
firms' objectives. The same approaches that have been developed (unfortunately
still under-developed) to measure increments to academic researchers' STHC could
be applied to industrial contexts. Indeed, such applications would in some way be
even more interesting because of the diverse needs and uses of STHC in industrial
settings.

5.2.4 Recommendation 4: Focus on Management of Large-Scale University-Based Research Centers

The few studies that have been conducted on management selection, strategy and
effectiveness in large-scale university centers with collaborative missions strongly

suggest the need for more such studies. The practice, widely if informally employed in universities and government policies, of elevating cooperative agreement P.I.'s to center directorships warrants study. The notion that someone will prove a good manager of enormous and multifaceted research institutions because they are talented and productive researchers is a dangerous one that is not born out by the limited research evidence (e.g. Rogers and Bozeman 2001; Corley et al. 2006; Philbin 2010; Cruz-Castro et al. 2012).

Complex university-based research centers do not at present ensure that those in charge of managing them have significant *managerial* knowledge and experience. While workshops on management and management handbooks (such as those provided by the National Science Foundation's Engineering Research Centers program) provide a useful supplement, they are no substitute for professional managerial training. More examination of the backgrounds, experiences and performance of academic managers of large, collaborative university research centers should shed some light on a variety of collaboration issues and, possibly, barriers.

5.2.5 Recommendation 5: Pursue Field Experiments

If the sub-literatures on research centers and team science (e.g., on NSF centers, NIH teams) can learn one important lesson from the literatures on inter-firm R&D alliances and industry R&D consortia, it is that they the need to transition from single-case "best practices" case study research to studying large numbers of organizations in quasi-experimental and field-experimental research. The changing goals from one research center or team of scientists to the next makes inductive case study valuable chiefly for retrospective evaluation rather than prospective team/center design and implementation. Thus, should a science team at the National Cancer Institute be a best practices exemplar for other NIH centers focused on alcoholism? Or for NSF centers focused on very different scientific and technical problems? Research on firms (e.g., Ettlie et al. 1984) shows that best practices conducive to collaboration in one context can have the opposite or null impact in other contexts due to difference in scientific and technical focus as well as many other contextual factors. The team science and research centers literatures claim to be in their "taxonomic" phases, thereby justifying a focus on the minutiae and idiosyncratic complexity of individual cases. We feel this conclusion is premature. Aggregate analysis and field experiment approaches borrowed from the firm literatures on collaborative R&D, especially studies focusing on governance mechanisms (e.g., structure, goal congruence, resource dependence, are at this point much more likely to prove productive than are single case studies or even multiple poorly integrated case studies. It is imperative at this point of research to avoid haphazard application of poorly verified "best practices."

5.2.6   Recommendation 6: More Impact-Focused Research; Integrated Collaborative Teams to Study Collaboration

If one compares the ratio of papers describing how boundary-spanning research occurs, including such topics as collaboration motivations, structuring, goals, and networking strategies, to papers identifying and explaining impacts, the clearly the scale is tilted toward answering "what?" questions rather than "so what?" questions. To be sure, the "what?" questions are important and, in at least some cases require additional attention. But if research is to be useful for other than explanatory theory purposes, then more attention should be paid to impact. This will not be easy. Let us consider why. It helps if we start by observing that there has been considerable progress mapping the impact of *individual* level collaborations on *individual* productivity (for an overview see Sonnenwald 2007; Bozeman et al. 2013). The reasons for progress are many but include the fact that citations and publication counts provide reasonably valid productivity measures.

Compared to simple collaborations, boundary-spanning collaborations are inordinately difficult to effectively and comprehensively evaluate because (1) almost all the organizations involved have their own multiple goals; (2) often there is goal conflict among the organizations party to collaboration, if not in absolute terms at least in order of priority. Impacts from complex boundary-spanning collaborations come in only two categories- those obviously difficult to measure and those that seem easy to measure but are not. For example, one of the most straightforward goals that some companies have for their collaborations is the development of technology that can result in products or process that can be brought to the market. It is easy enough to understand the appeal of this goal. However, since companies usually have a great many inputs pertaining to product development (in-house R&D, strategic choices, production capacity, manufacturing process and skills, marketing analysis and strategy, to name a few) it is nearly impossible to isolate the specific contribution of particular collaborative activities to the realization of what is an easily understood goal.

Most studies examining impacts of complex boundary-spanning collaborations rely on count data (number of patents) or self reports through interviews and questionnaires. Such studies have known but also quite significant problems. The problem with patents is that they are usually a surrogate measure and sometimes not a particularly useful one. Self-reports are often useful, but they are most useful when one is confident that the person providing the report has the full knowledge required for a valid report. Sometimes they do. So, for example, if a researcher is asked "have you ever collaborated with an employee of a government lab?" the subject has full knowledge needed to provide a potentially valid report. But if one asks a Vice President for Research, "did the collaboration in University-Industry Center result in a product brought to market" then one faces a question that only seems to have a simple answer. The problem with self-reports is that they are notoriously unreliable when pertaining to information characterizing a whole organization and its activities (much less multiple organizations and multiple loosely-connected technically-relevant activities).

If one wanted more and better research on the impacts of complex boundary-spanning collaborations, how might one proceed? A key point is to rely on more than a single informant. In fact, it would be useful to obtain data from several persons, with different roles and perspectives, from various components of the organization and to do this within all of the participating organizations. A second point is to avoid snapshots. For obvious reasons (lack of sufficient funding and lack of time) most of the studies drawing upon unique information from people working in organizations are cross-sectional and examine a narrow time band.

From a practical standpoint how could these limitations be addressed? One easy answer is "more funding," but without more funding perhaps more integrated, collaborative team research- a reflexive approach to understanding collaboration and teams. In the physical and natural science there has been considerable progress made owing to policy-makers *requirements* that center affiliates or program affiliates work together on integrated research tasks. We can perhaps assume that any funding announcement requiring teams to focus integrated research, over several years, to study multiple complex collaborations would not fail due to lack of proposals.

References

Aerts, K., & Schmidt, T. (2008). Two for the price of one? Additionality effects of R&D subsidies: A comparison between Flanders and Germany. *Research Policy, 37*(5), 806–822.

Agarwal, R., & Ohyama, A. (2013). Industry or academia, basic or applied? Career choices and earnings trajectories of scientists. *Management Science, 59*(4), 950–970.

Agrawal, A., & Henderson, R. (2002). Putting patents in context: Exploring knowledge transfer from MIT. *Management Science, 48*(1), 44–60.

Allchin, D. (2003). Scientific myth-conceptions. *Science Education, 87*(3), 329–351.

Ambos, T. C., Mäkelä, K., Birkinshaw, J., & D'Este, P. (2008). When does university research get commercialized? Creating ambidexterity in research institutions. *Journal of Management Studies, 45*(8), 1424–1447.

Amick, D. J. (1973). The scientist's mission and professional involvement: A quantitative revisitation to the sociology of science. *Social Science Research, 2*(3), 293–306.

Andereggen, S., Zoller, F. A., & Boutellier, R. (2013). Sharing research equipment to bridge intra-organizational boundaries: The cases of novartis and ETH zurich. *Research-Technology Management, 56*(1), 49–57.

Audretsch, D. B., Bozeman, B., Combs, K. L., Feldman, M., Link, A. N., Siegel, D. S., & Wessner, C. (2002). The economics of science and technology. *The Journal of Technology Transfer, 27*(2), 155–203.

Azoulay, P., Dng, W., & Stuart, T. (2009). The impact of academic patenting on the rate, quality and direction of (Public) research output. *The Journal of Industrial Economics, 57*(4), 637–676.

Backes, P. G., Tso, K. S., Norris, J. S., & Steinke, R. (2002). Group collaboration for Mars rover mission operations. In Robotics and Automation (Ed.), *Proceedings. ICRA'02. IEEE International Conference on* (Vol. 3, pp. 3148–3154). IEEE.

Baldini, N. (2008) Negative effects of university patenting: Myths and grounded evidence. *Scientometrics, 75*(2), 289–311.

Banal-Estañol, A., & Macho-Stadler, I. (2010). Scientific and commercial incentives in R&D: Research versus development? *Journal of Economics & Management Strategy, 19*(1), 185–221.

Bekelman, J. E., Li, Y., & Gross, C. P. (2003). Scope and impact of financial conflicts of interest in biomedical research. *JAMA: The Journal of the American Medical Association, 289*(4), 454–465.

Belderbos, R., Carree, M., & Lokshin, B. (2004). Cooperative R&D and firm performance. *Research Policy, 33*(10), 1477–1492.

Belderbos, R., Carree, M., & Lokshin, B. (2006). Complementarity in R&D cooperation strategies. *Review of Industrial Organization, 28*(4), 401–426.

Berchicci, L. (2013). Towards an open R&D system: Internal R&D investment, external knowledge acquisition and innovative performance. *Research Policy, 42*(1), 117–127.

Bercovitz, J. E., & Feldman, M. P. (2007). Fishing upstream: Firm innovation strategy and university research alliances. *Research Policy, 36*(7), 930–948.

Besancenot, D., Huynh, K., & Vranceanu, R. (2011). A matching model of the academic publication market. *Journal of Institutional and Theoretical Economics (JITE)*, *167*(4), 708–725.

Blackman, S. H., & McNeill, R. M. (1997). Alternative dispute resolution in commercial intellectual property disputes. *American University Law Review*, *47*, 1709.

Blumenthal, D., Campbell, E. G., Anderson, M. S., Causino, N., & Louis, K. S. (1997). Withholding research results in academic life science. *JAMA: The Journal of the American Medical Association*, *277*(15), 1224–1228.

Blumenthal, D., Campbell, E. G., Causino, N., & Louis, K. S. (1996). Participation of life-science faculty in research relationships with industry. *New England Journal of Medicine*, *335*(23), 1734–1739.

Boardman, P. C. (2009). Government centrality to university–industry interactions: University research centers and the industry involvement of academic researchers. *Research Policy*, *38*(10), 1505–1516.

Bojanowski, M., Corten, R., & Westbrock, B. (2012). The structure and dynamics of the global network of inter-firm R&D partnerships 1989–2002. *The Journal of Technology Transfer*, *37*(6), 967–987.

Bordons, M., Zulueta, M. A., Romero, F., & Barrigón, S. (1999). Measuring interdisciplinary collaboration within a university: The effects of the multidisciplinary research programme. *Scientometrics*, *46*(3), 383–398.

Boring, E. G. (1950). Great men and scientific progress. *Proceedings of the American Philosophical Society*, *94*(4), 339–351.

Bort, S., & Kieser, A. (2011). Fashion in organization theory: An empirical analysis of the diffusion of theoretical concepts. *Organization Studies*, *32*(5), 655–681.

Bozeman, B., & Boardman, C. (2004). The NSF engineering research centers and the university–industry research revolution: A brief history featuring an interview with Erich Bloch. *The Journal of Technology Transfer*, *29*(3–4), 365–375.

Bozeman, B., & Boardman, C. (2013). Academic faculty in university research centers: Neither capitalism's slaves nor teaching fugitives. *The Journal of Higher Education*, *84*(1), 88–120.

Bozeman, B., & Corley, E. (2004). Scientists' collaboration strategies: Implications for scientific and technical human capital. *Research Policy*, *33*(4), 599–616.

Bozeman, B., & Papadakis, M. (1995). Company interactions with federal laboratories: What they do and why they do it. *The Journal of Technology Transfer*, *20*(3–4), 64–74.

Bozeman, B., & Wittmer, D. (2001). Technical roles and success of US federal laboratory-industry partnerships. *Science and Public Policy*, *28*(3), 169–178.

Bozeman, B., Dietz, J. S., & Gaughan, M. (2001). Scientific and technical human capital: An alternative model for research evaluation. *International Journal of Technology Management*, *22*(7), 716–740.

Bozeman, B., Fay, D., & Slade, C. P. (2013). Research collaboration in universities and academic entrepreneurship: The-state-of-the-art. *The Journal of Technology Transfer*, *38*(1), 1–67.

Breschi, S., Lissoni, F., & Montobbio, F. (2007). The scientific productivity of academic inventors: New evidence from Italian data. *Economics of Innovation and New Technology*, *16*(2), 101–118.

Caloghirou, Y., Tsakanikas, A., & Vonortas, N. S. (2001). University-industry cooperation in the context of the European framework programmes. *The Journal of Technology Transfer*, *26*(1), 153–161.

Campbell, E. G., Clarridge, B. R., Gokhale, M., Birenbaum, L., Hilgartner, S., Holtzman, N. A., & Blumenthal, D. (2002). Data withholding in academic genetics. *JAMA: The Journal of the American Medical Association*, *287*(4), 473–480.

Carayol, N., & Matt, M. (2004). Does research organization influence academic production? Laboratory level evidence from a large European university. *Research Policy*, *33*, 1081–1102.

Cohen, W. M., Florida, R., Randazzese, L., & Walsh, J. (1998). Industry and the academy: Uneasy partners in the cause of technological advance. *Challenges to Research Universities*, *171*, 200.

Contractor, F.J., & Lorange, P., (1988). Why should firms cooperate? The strategy and economics basis for cooperative ventures. In F. J. Contractor & P. Lorange (Eds.), *Cooperative strategies in international business* (pp. 3–30). Lexington: Lexington Books.

Cooksy, L. J., Gill, P., & Kelly, P. A. (2001). The program logic model as an integrative framework for a multimethod evaluation. *Evaluation and Program Planning, 24*(2), 119–128.

Corley, E. A., Boardman, P. C., & Bozeman, B. (2006). Design and the management of multi-institutional research collaborations: Theoretical implications from two case studies. *Research Policy, 35*(7), 975–993.

Corolleur, C. D., Carrere, M., & Mangematin, V. (2004). Turning scientific and technological human capital into economic capital: The experience of biotech start-ups in France. *Research Policy, 33*(4), 631–642.

Crane, D. (1969). Fashion in science: Does it exist? *Social Problems, 16*, 433–441.

Cronin, B. (1984). *The citation process: The role and significance of citations in scientific communication.* London: Taylor Graham.

Cruz-Castro, L., Sanz-Menéndez, L., & Martínez, C. (2012). Research centers in transition: Patterns of convergence and diversity. *The Journal of Technology Transfer, 37*(1), 18–42.

Davenport, S. (2004). Panic and panacea: Brain drain and science and technology human capital policy. *Research Policy, 33*(4), 617–630.

Davenport, S., Davies, J., & Grimes, C. (1998). Collaborative research programmes: Building trust from difference. *Technovation, 19*(1), 31–40.

De Stefano, D., Fuccella, V., Vitale, M. P., & Zaccarin, S. (2013). The use of different data sources in the analysis of co-authorship networks and scientific performance *Social Networks, 35*, 370–381.

Demirkan, I., & Deeds, D. L. (2013). The evolution of research collaboration networks and their impact on firm innovation output. *Technology, Innovation, Entrepreneurship and Competitive Strategy, 13*, 67–95.

D'Este, P., Guy, F., & Iammarino, S. (2013). Shaping the formation of university–industry research collaborations: What type of proximity does really matter? *Journal of Economic Geography, 13*(4), 537–558.

Dietz, J. S., & Bozeman, B. (2005). Academic careers, patents, and productivity: Industry experience as scientific and technical human capital. *Research Policy, 34*(3), 349–367.

DiMaggio, P. (1998). The new institutionalisms: Avenues of collaboration. *Journal of Institutional and Theoretical Economics (JITE)/Zeitschrift für die gesamte Staatswissenschaft, 154*(4), 696–705.

Duysters, G., & Hagedoorn, J. (1996). Internationalization of corporate technology through strategic partnering: An empirical investigation. *Research Policy 25*, 1–12.

Eden, L., Hitt, M., Ireland, R. D., Garrett, R., & Li, D. (2011). Governance in multilateral R&D alliances. *Organization Science, 23*(4), 1–20.

Edler, J., Fier, H., & Grimpe, C. (2011). International scientist mobility and the locus of knowledge and technology transfer. *Research Policy, 40*(6), 791–805.

Eisenhardt, K. M., & Schoonhoven, C. B. (1996). Resource-based view of strategic alliance formation: Strategic and social effects in entrepreneurial firms. *Organization Science 7*, 136–150.

Faems, D., Janssens, M., Bouwen, R., & Van Looy, B. (2006). Governing explorative R&D alliances: Searching for effective strategies. *Management Revue. The International Review of Management Studies, 17*(1), 9–29.

Feller, I., & Roessner, D. (1995). What does industry expect from university partnerships? Congress wants to see bottom-line results from industry/government programs, but that's not what the participating companies are seeking. *Issues in Science and Technology, 12*(1) 80–84.

Feller, I., Ailes, C. P., & Roessner, J. D. (2002). Impacts of research universities on technological innovation in industry: Evidence from engineering research centers. *Research Policy, 31*(3), 457–474.

Fontana, R., Geuna, A., & Matt, M. (2006). Factors affecting university-industry R&D projects: The importance of searching, screening and signalling. *Research Policy, 35*, 309–323.

Fox, M. F., & Stephan, P. E. (2001). Careers of young scientists: Preferences, prospects and realities by gender and field. *Social Studies of Science, 31*(1), 109–122.

Freeman, C., & Hagedoorn, J. (1994). Catching up or falling behind: Patterns in international interfirm partnering. *World Development 22*, 771–780.

Fuhrmann, C. N., Halme, D. G., O'Sullivan, P. S., & Lindstaedt, B. (2011). Improving graduate education to support a branching career pipeline: Recommendations based on a survey of doctoral students in the basic biomedical sciences. *CBE-Life Sciences Education, 10*(3), 239–249.

Fulton, O., & Trow, M. (1974). Research activity in American higher education. *Sociology of Education, 47*, 29–73.

Garcia, R., & Calantone, R. (2002). A critical look at technological innovation typology and innovativeness terminology: A literature review. *Journal of Product Innovation Management, 19*(2), 110–132.

García-Canal, E., Valdés-Llaneza, A., & Sánchez-Lorda, P. (2014). Contractual form in repeated alliances with the same partner: The role of inter-organizational routines. *Scandinavian Journal of Management, 30*(1), 51–64.

Garfield, E. (1979) *Citation indexing: Its theory and applications in science, technology and the humanities.* New York: Wiley.

Gassmann, O., Enkel, E., & Chesbrough, H. (2010). The future of open innovation. *R&D Management, 40*(3), 213–221.

Gaughan, M., & Robin, S. (2004). National science training policy and early scientific careers in France and the United States. *Research Policy, 33*(4), 569–581.

Geuna, A., & Nesta, L. (2006). University patenting and its effects on academic research: The emerging European evidence. *Research Policy, 35*, 790–807.

Gluck, M., Blumenthal, D., & Stoto, M. A. (1987). University-industry relationships in the life sciences: Implications for students and post-doctoral fellows. *Research Policy, 16*, 327–336.

Godin, B. (2006). On the origins of bibliometrics. *Scientometrics, 68*(1), 109–133.

Gomes-Casseres, B. (1996). *The alliance revolution: The new shape of business rivalry.* Cambridge: Harvard University Press.

Gray, D. O., & Steenhuis, H. J. (2003). Quantifying the benefits of participating in an industry university research center: An examination of research cost avoidance. *Scientometrics, 58*(2), 281–300.

Gulati, R. (1995). Does familiarity breed trust? The implications of repeated ties for contractual choice in alliances. *Academy of Management Journal, 38*(1), 85–112.

Gulbrandsen, M., & Smeby, J. C. (2005). Industry funding and university professors' research performance. *Research Policy, 34*(6), 932–950.

Hagedoorn, J. (2002). Inter-firm R&D partnerships: An overview of major trends and patterns since 1960. *Research Policy, 31*(4), 477–492.

Hagedoorn, J., & Wang, N. (2012). Is there complementarity or substitutability between internal and external R&D strategies? *Research Policy, 41*(6), 1072–1083.

Hagedoorn, J. (1990). Organizational modes of inter-firm cooperation and technology transfer. *Technovation, 10*, 17–30.

Hagedoorn, J. (1993). Understanding the rationale of strategic technology partnering: Inter-organizational modes of cooperation and sectoral differences. *Strategic Management Journal 14*, 371–385.

Hagedoorn, J. (1996). Trends and patterns in strategic technology partnering since the early seventies. *Review of Industrial Organization, 11*, 601–616.

Hagedoorn, J., Link, A. N., & Vonortas, N. S. (2000). Research partnerships. *Research Policy, 29*(4), 567–586.

Hagedoorn, J., Link, A. N., & Vonortas, N. S. (2000). Research partnerships. *Research Policy, 29*(4), 567–586.

Hagedoorn, J., & Narula, R., (1996). Choosing organizational modes of strategic technology partnering: International and sectoral differences. *Journal of International Business Studies, 27*, 265–284.

Hagedoorn, J., & Schakenraad, J. (1993). A comparison of private and subsidized inter-firm linkages in the European IT industry. *Journal of Common Market Studies, 31*, 373–390.

Hall, K. L., Stokols, D., Moser, R. P., Taylor, B. K., Thornquist, M. D., Nebeling, L. C., & Jeffery, R. W. (2008). The collaboration readiness of transdisciplinary research teams and centers: Findings from the National Cancer Institute's TREC year-one evaluation study. *American Journal of Preventive Medicine, 35*(2), S161–S172.

Harrigan, K. R., & Newman, W. H. (1990). Bases of interorganization co-operation: Propensity, power, persistence. *Journal of Management Studies, 27*, 417–434.

Haustein, S., Tunger, D., Heinrichs, G., & Baelz, G. (2011). Reasons for and developments in international scientific collaboration: Does an Asia–Pacific research area exist from a bibliometric point of view? *Scientometrics, 86*(3), 727–746.

Hayton, J. C., Sehili, S., & Scarpello, V. (2013). Why do firms join cooperative research centers? An empirical examination of firm, industry, and environmental antecedents. In *Cooperative Research Centers and Technical Innovation* (pp. 37–57). New York: Springer.

Heller, M. A., & Eisenberg, R. S. (1998), Can patents deter innovation? The anticommons in biomedical research. *Science, 280*, 698–701.

Henderson, R., Jaffe, A. B., & Trajtenberg, M. (1998). Universities as a source of commercial technology: A detailed analysis of university patenting, 1965–1988. *Review of Economics and Statistics, 80*(1), 119–127.

Hernández-Espallardo, M., Sánchez-Pérez, M., & Segovia-López, C. (2011). Exploitation-and exploration-based innovations: The role of knowledge in inter-firm relationships with distributors. *Technovation, 31*(5), 203–215.

Hertzfeld, H., Link A., & Vonortas N. (2006). Intellectual property protection mechanisms in research partnerships. *Research Policy, 35*, 825–838.

Hessels, L. K., & Van Lente, H. (2008). Re-thinking new knowledge production: A literature review and a research agenda. *Research Policy, 37*(4), 740–760.

Hicks, D., Breitzman, T., Olivastro, D., & Hamilton, K. (2001). The changing composition of innovative activity in the US—A portrait based on patent analysis. *Research Policy, 30*(4), 681–703.

Hicks, D., & Kimberly H. (1999). "Does university-industry collaboration adversely affect university research?" Issues in Science and Technology.

Hoegl, M., & Proserpio, L. (2004). Team member proximity and teamwork in innovative projects. *Research Policy, 33*(8), 1153–1165.

Hoekman, J., Frenken, K., & Tijssen, R. J. (2010). Research collaboration at a distance: Changing spatial patterns of scientific collaboration within Europe. *Research Policy, 39*(5), 662–673.

Hood, W. W., & Wilson, C. S. (2001). The literature of bibliometrics, scientometrics, and informetrics. *Scientometrics, 52*(2), 291–314.

Hottenrott, H., & Thorwarth, S. (2011). Industry funding of university research and scientific productivity. *Kyklos, 64*(4), 534–555.

Hovenkamp, H., Janis, M., & Lemley, M. A. (2002). Anticompetitive settlement of intellectual property disputes. *Minnesota Law Review, 87*, 1719.

Huang, K. G., & Murray, F. E. (2009). Does patent strategy shape the long-run supply of public knowledge? Evidence from human genetics. *Academy of Management Journal, 52*(6), 1193–1221.

Hula, K. W. (1999). *Lobbying together: Interest group coalitions in legislative politics*. Washington: Georgetown University Press.

Hummon, N. P., & Dereian, P. (1989). Connectivity in a citation network: The development of DNA theory. *Social Networks, 11*(1), 39–63.

Jonkers, K., & Cruz-Castro, L. (2013). Research upon return: The effect of international mobility on scientific ties, production and impact. *Research Policy, 42*(8), 1366–1377.

Jonkers, K., & Tijssen, R. (2008). Chinese researchers returning home: Impacts of international mobility on research collaboration and scientific productivity. *Scientometrics, 77*(2), 309–333.

Julian, D. (1997). The utilization of the logic model as a system level planning and evaluation device. *Evaluation and Program Planning, 20*(3), 251–257.

Katz, J. S., & Hicks, D. (1997). How much is a collaboration worth? A calibrated bibliometric model. *Scientometrics, 40*(3), 541–554.

Kellogg, F. (2001). Logic model development guide: Logic models to bring together planning, evaluation & action. Battle Creek: W.K. Kellogg Foundation.

Kenney, M., & Richard Goe, W. (2004). The role of social embeddedness in professorial entrepreneurship: A comparison of electrical engineering and computer science at UC Berkeley and Stanford. *Research Policy, 33*(5), 691–707.

Kleinman, D. L. (2003). *Impure cultures: University biology and the world of commerce.* Madison: University of Wisconsin Press.

Knoben, J., & Oerlemans, L. A. (2006). Proximity and inter-organizational collaboration: A literature review. *International Journal of Management Reviews, 8*(2), 71–89.

Knowlton, L. W., & Phillips, C. C. (2012). The logic model guidebook: Better strategies for great results. Thousand Oaks: Sage.

Krogmann, Y., Riedel, N., & Schwalbe, U. (2013). Inter-firm R&D networks in pharmaceutical biotechnology: What determines firm's centrality-based partnering capability (No. 75–2013). FZID Discussion Paper.

Lawson, B., & Potter, A. (2012). Determinants of knowledge transfer in inter-firm new product development projects. *International Journal of Operations & Production Management, 32*(10), 1228–1247.

Lee, C. Y., & Chang, S. C. (2012). How do contextual factors affect a firm's R&D alliance performance? A study of biotech industry. Management of innovation and technology (ICMIT), 2012 IEEE International Conference on (pp. 62–67). IEEE.

Lee, S., & Bozeman, B. (2005). The impact of research collaboration on scientific productivity. *Social Studies of Science, 35*(5), 673.

Lee, Y. S. (2000). The sustainability of university-industry research collaboration: An empirical assessment. *The Journal of Technology Transfer, 25*(2), 111–133.

Levy, R., Roux, P., & Wolff, S. (2009). An analysis of science–industry collaborative patterns in a large European university. *The Journal of Technology Transfer, 34*(1), 1–23.

Lightman, B. (Ed.). (2008). *Victorian science in context.* Chicago: University of Chicago Press.

Lin, C., Wu, Y. J., Chang, C., Wang, W., & Lee, C. Y. (2012). The alliance innovation performance of R&D alliances—the absorptive capacity perspective. *Technovation, 32*(5), 282–292.

Lin, M. W., & Bozeman, B. (2006). Researchers' industry experience and productivity in university–industry research centers: A "scientific and technical human capital" explanation. *The Journal of Technology Transfer, 31*(2), 269–290.

Link, A. N., & Long, J. E. (1981). The simple economics of basic scientific research: A test of Nelson's diversification hypothesis. *The Journal of Industrial Economics, 30*(1), 105–109.

Link, A. N., Siegel, D. S., & Bozeman, B. (2007). An empirical analysis of the propensity of academics to engage in informal university technology transfer. *Industrial and Corporate Change, 16*(4), 641–655.

Lo, B., & Field, M. J. (Eds.). (2009). *Conflict of interest in medical research, education, and practice.* Washington: National Academies Press.

Lööf, H., & Broström, A. (2008). Does knowledge diffusion between university and industry increase innovativeness? *The Journal of Technology Transfer, 33*(1), 73–90.

Mandel, G. N. (2011). To promote the creative process: Intellectual property law and the psychology of creativity. *Notre Dame Law Review, 86*, 1999.

Mansfield, E. (1995). Academic research underlying industrial innovations: Sources, characteristics, and financing. *The Review of Economics and Statistics, 77*(1), 55–65.

Mariti, P., & Smiley, R. H. (1983). Co-operative agreements and the organization of industry. *Journal of Industrial Economics, 31*, 3437–3451.

Martin, H., & Diego, N. (2009). Where should we submit our manuscript? An analysis of journal submission strategies. *The BE Journal of Economic Analysis & Policy, 9*(1), 1–28.

Mathieu, J., Maynard, M. T., Rapp, T., & Gilson, L. (2008). Team effectiveness 1997–2007: A review of recent advancements and a glimpse into the future. *Journal of Management, 34*(3), 410–476.

McLaughlin, J., & Jordan, G. (1999). Logic models: A tool for telling your program's performance story. *Evaluating and Program Planning, 22*, 65–72.

Menzies, M. B. (2012). Researching scientific entrepreneurship in New Zealand. *Science and Public Policy, 39*(1), 39–59.

Meyer, M. (2006). Are patenting scientists the better scholars? An exploratory comparison of inventor authors with their non-inventing peers in nanoscience and technology. *Research Policy, 35*, 1646–1662.

Miller, G. A. (1967). Professionals in bureaucracy: Alienation among industrial scientists and engineers. *American Sociological Review, 32*(5), 755–768.

Miotti, L., & Sachwald, F. (2003). Co-operative R&D: Why and with whom? An integrated framework of analysis. *Research Policy, 32*(8), 1481–1499.

Murray, F. (2010). The oncomouse that roared: Hybrid exchange strategies as a source of distinction at the boundary of overlapping institutions. *American Journal of Sociology, 116*(2), 341–388.

Murray, F., & Stern, S. (2007). Do formal intellectual property rights hinder the free flow of scientific knowledge? An empirical test of the anti-commons hypothesis. *Journal of Economic Behavior & Organization, 63*(4), 648–687.

Muscio, A., & Pozzali, A. (2012). The effects of cognitive distance in university-industry collaborations: Some evidence from Italian universities. *The Journal of Technology Transfer, 38*(4), 1–23.

Narula, R., & Dunning, J. H. (1998). Explaining international R&D alliances and the role of governments. *International Business Review, 7*(4), 377–397.

Nedeva, M., Georghiou, L., & Halfpenny, P. (1999). Benefactors or beneficiary—The role of industry in the support of university research equipment. *The Journal of Technology Transfer, 24*(2–3), 139–147.

Nelson, R. R. (1982). The role of knowledge in R&D efficiency. *The Quarterly Journal of Economics, 97*(3), 453–470.

Onel, S., Zeid, A., & Kamarthi, S. (2011). The structure and analysis of nanotechnology co-author and citation networks. *Scientometrics, 89*(1), 119–138.

Osborn, R. N., & Baughn, C. C., (1990). Forms of interorganizational governance for multinational alliances. *Academy of Management Journal, 33*, 503–519.

Oxley, J. E., & Sampson, R. C. (2004). The scope and governance of international R&D alliances. *Strategic Management Journal, 25*(8–9), 723–749.

Ozman, M. (2009). Inter-firm networks and innovation: A survey of literature. *Economic of Innovation and New Technology, 18*(1), 39–67.

panel dataset of 3,862 academic life scientists.

Panico, C. (2011). On the contractual governance of research collaborations: Allocating control and intellectual property rights in the shadow of potential termination. *Research Policy, 40*(10), 1403–1411.

Paruchuri, S. (2012). Inventor sourcing of alliance partners' knowledge: Role of intra-firm inventor networks. *Academy of management proceedings* (Vol. 2012, No. 1, pp. 1–1). Academy of Management.

Perkmann, M., Neely, A., & Walsh, K. (2011). How should firms evaluate success in university–industry alliances? A performance measurement system. *R&D Management, 41*(2), 202–216.

Perkmann, M., Tartari V., McKelvey M., Autio E., Broström A., D'Este P., Fini R., Geuna A., Grimaldi R., Hughes A., Krabel S., Kitson M., Llerena P., Lissoni F., Salter A., & Sobrero M. (2013). Academic engagement and commercialisation: A review of the literature on university–industry relations. *Research Policy, 42*(2), 423–442.

Persson, O. (2010). Are highly cited papers more international? *Scientometrics, 83*(2), 397–401.

Phelps, A. F., & Reddy, M. (2009). The influence of boundary objects on group collaboration in construction project teams. Proceedings of the ACM 2009 international conference on Supporting group work (pp. 125–128). ACM.

Philbin, S. P. (2010). Developing and managing university-industry research collaborations through a process methodology/industrial sector approach. *Journal of Research Administration, 41*(3), 51–68.

Ponomariov, B. (2013). Government-sponsored university-industry collaboration and the production of nanotechnology patents in US universities. *The Journal of Technology Transfer, 38*(6), 1–19.

Ponomariov, B. L., & Boardman, P. C. (2010). Influencing scientists' collaboration and productivity patterns through new institutions: University research centers and scientific and technical human capital. *Research Policy, 39*(5), 613–624.

Porter, A. L., & Rafols, I. (2009). Is science becoming more interdisciplinary? Measuring and mapping six research fields over time. *Scientometrics, 81*(3), 719–745.

Porter, A. L., Roessner, J. D., Cohen, A. S., & Perreault, M. (2006). Interdisciplinary research: Meaning, metrics and nurture. *Research Evaluation, 15*(3), 187–195.

Powell, W. W., Koput, K. W., & Smith-Doerr, L. (1996). Interorganizational collaboration and the locus of innovation: Networks of learning in biotechnology. *Administrative science quarterly,* 116–145.

Pravdić, N., & Oluić-Vuković, V. (1986). Dual approach to multiple authorship in the study of collaboration/scientific output relationship. *Scientometrics, 10*(5), 259–280.

Qin, J., Lancaster, F. W., & Allen, B. (1997). Types and levels of collaboration in interdisciplinary research in the sciences. *Journal of the American Society for information Science, 48*(10), 893–916.

Reagan, M. D. (1967). Basic and applied research: A meaningful distinction? *Science, 155*(3768), 1383–1386.

Reuer, J. J., & Zollo, M. (2005). Termination outcomes of research alliances. *Research Policy, 34*(1), 101–115.

van Rijnsoever, F. J., & Hessels, L. K. (2011). Factors associated with disciplinary and interdisciplinary research collaboration. *Research Policy, 40*(3), 463–472.

Roach, M., & Sauermann, H. (2010). A taste for science? PhD scientists' academic orientation and self-selection into research careers in industry. *Research Policy, 39*(3), 422–434.

Rogers, J. D. (2012). Research centers as agents of change in the contemporary academic landscape: Their role and impact in HBCU, EPSCoR, and majority universities. *Research Evaluation, 21*(1), 15–32.

Rogers, J. D., & Bozeman, B. (1997). Basic research and the success of federal lab-industry partnerships. *The Journal of Technology Transfer, 22*(3), 37–47.

Rogers, J. D., & Bozeman, B. (2001). "Knowledge value alliances": An alternative to the R&D project focus in evaluation. *Science, Technology & Human Values, 26*(1), 23–55.

Saavedra, P., & Bozeman, B. (2004). The "gradient effect" in federal laboratory-industry technology transfer partnerships. *Policy Studies Journal, 32*(2), 235–252.

Sakakibara, M. (1997). Heterogeneity of firm capabilities and cooperative research and development: An empirical examination of motives. *Strategic Management Journal, 18*(S1), 143–164.

Sampson, R. C. (2004). Organizational choice in R&D alliances: Knowledge-based and transaction cost perspectives. *Managerial and Decision Economics, 25*(6–7), 421–436.

Sampson, R. C. (2004). The cost of misaligned governance in R&D alliances. *Journal of Law, Economics, and Organization, 20*(2), 484–526.

Sampson, R. C. (2005). Experience effects and collaborative returns in R&D alliances. *Strategic Management Journal, 26*(11), 1009–1031.

Sampson, R. C. (2007). R&D alliances and firm performance: The impact of technological diversity and alliance organization on innovation. *Academy of Management Journal, 50*(2), 364–386.

Sauermann, H., & Roach, M. (2014). Not all scientists pay to be scientists: PhDs' preferences for publishing in industrial employment. *Research Policy, 43*(1), 32–47.

Sauermann, H., & Stephan, P. (2013). Conflicting logics? A multidimensional view of industrial and academic science. *Organization Science, 24*(3), 889–909.

Schmoch, U. (1999). Interaction of universities and industrial enterprises in Germany and the United States-a comparison. *Industry and Innovation, 6*(1), 51–68.

Schummer, J. (2004). Multidisciplinarity, interdisciplinarity, and patterns of research collaboration in nanoscience and nanotechnology. *Scientometrics*, *59*(3), 425–465.

Shaw, J. D., Zhu, J., Duffy, M. K., Scott, K. L., Shih, H. A., & Susanto, E. (2011). A contingency model of conflict and team effectiveness. *Journal of Applied Psychology*, *96*(2), 391.

Shrivastava, P., & Mitroff, I. I. (1984). Enhancing organizational research utilization: The role of decision makers' assumptions. *Academy of Management Review*, *9*(1), 18–26.

Siegel, D., Waldman D., & Link A. (2003). Assessing the impact of organizational practices on the productivity of university technology transfer offices: An exploratory study. *Research Policy*, *32*, 27–48.

Slaughter, S., & Leslie, L. L. (1997). Academic capitalism: politics, policies, and the entrepreneurial university. Baltimore: Johns Hopkins University Press.

Slaughter, S., T. Campbell, M. H. Folleman, & E. Morgan (2002) The "traffic" in graduate students: Graduate students as tokens of exchange between academe and industry. *Science, Technology and Human Values*, *27*(2), 282–313.

Sonnenwald, D. H. (2007). Scientific collaboration. *Annual Review of Information Science and Technology*, *41*(1), 643–681.

Stuart, T. E., & Ding, W. W. (2006). When do scientists become entrepreneurs? The social structural antecedents of commercial activity in the academic life sciences1. *American Journal of Sociology*, *112*(1), 97–144.

Susser, M., & Yankauer, A. (1993). Prior, duplicate, repetitive, fragmented, and redundant publication and editorial decisions. *American Journal of Public Health*, *83*(6), 792–793.

Tacke, O. (2011). Open science 2.0: How research and education can benefit from open innovation and Web 2.0. In *On collective intelligence* (pp. 37–48).Berlin: Springer.

Tartari, V., & Breschi, S. (2011). Set them free: Scientists' evaluations of benefits and costs of university-industry research collaboration. *Industrial and Corporate Change*, *21*(5), 1117–1147.

Tartari, V., Salter, A., & D'Este, P. (2012). Crossing the Rubicon: Exploring the factors that shape academics' perceptions of the barriers to working with industry. *Cambridge Journal of Economics*, *36*(3), 655–677.

Thursby, J. G., & Thursby, M. C. (2002). Who is selling to the ivory tower? Sources of growth in university licensing. *Management Science*, *48*, 90–104.

Thursby J. G., & Thursby, M. C. (2003). Industry/University licensing: Characteristics, concerns and issues from the perspective of the buyer. *Journal of Technology Transfer*, *28*, 207–213.

Thursby J. G., & Thursby, M. C. (2004). Are faculty critical? Their role in university-industry licensing. *Contemporary Economic Policy*, *22*(4), 162–170.

Toole, A. A., & Czarnitzki, D. (2010). Commercializing science: Is there a university "brain drain" from academic entrepreneurship? *Management Science*, *56*(9), 1599–1614.

Van de Ven, A. H., Ganco, M., & Hinings, C. R. (2013). Returning to the frontier of contingency theory of organizational and institutional designs. *The Academy of Management Annals*, *7*(1), 393–440.

Van Looy, B., Callaert, J., Debackere, K. (2006). Publication and patent behaviour of academic researchers: Conflicting, reinforcing or merely co-existing? *Research Policy*, *35*, 596–608.

Van Looy, B., Ranga, M., Callaert, J., Debackere, K., & Zimmermann, E. (2004). Combining entrepreneurial and scientific performance in academia: Towards a compounded and reciprocal Matthew-effect? *Research Policy*, *33*(3), 425–441.

Vonortas, N. S. (1997). Research joint ventures in the US. *Research Policy*, *26*(4), 577–595.

Wallerstein, N., & Duran, B. (2010). Community-based participatory research contributions to intervention research: The intersection of science and practice to improve health equity. *American Journal of Public Health*, *100*(S1), S40–S46.

Webb, N. M. (1995). Group collaboration in assessment: Multiple objectives, processes, and outcomes. *Educational Evaluation and Policy Analysis*, *17*(2), 239–261.

Whissell, C. (1999). Linguistic complexity of abstracts and titles in highly cited journals. *Perceptual and Motor Skills*, *88*(1), 76–86.

Woolley, R., Turpin, T., Marceau, J., & Hill, S. (2008). Mobility matters: Research training and network building in science. *Comparative Technology Transfer and Society*, *6*(3), 159–184.

Wuchty, S., Jones, B. F., & Uzzi, B. (2007). The increasing dominance of teams in production of knowledge. *Science, 316*(5827), 1036.

Ynalvez, M. A., & Shrum, W. M. (2011). Professional networks, scientific collaboration, and publication productivity in resource-constrained research institutions in a developing country. *Research Policy, 40*(2), 204–216.

van Zeebroeck, N., de la Potterie, B. V. P., & Guellec, D. (2008). Patents and academic research: A state of the art. *Journal of Intellectual Capital, 9*(2), 246–263.

Zeng, S. X., Xie, X. M., & Tam, C. M. (2010). Relationship between cooperation networks and innovation performance of SMEs. *Technovation, 30*(3), 181–194.

Zhou, W., Zou, Y., Zhu, Y., Fei, S., & Lu, X. (2012, September). Wiki lab: A collaboration-oriented scitentific research platform. Electronic System-Integration Technology Conference (ESTC), 4th (pp. 411–414). IEEE.

Ziman, J. (1981). What are the options? Social determinants of personal research plants. *Minerva, 19*(1), 1–42.